A History of Medical
Bacteriology and Immunology

A History of Medical Bacteriology and Immunology

W. D. FOSTER
M.D. Cantab., M.R.C. Path. Reader in Microbiology
King's College Hospital Medical School

WILLIAM HEINEMANN MEDICAL BOOKS LTD.

First published 1970
© W. D. Foster, 1970
ISBN 0 433 10690 5

Printed in Great Britain by Cox & Wyman Ltd.
London, Fakenham and Reading

Contents

Foreword

Not many have written histories of medical bacteriology. Whatever the reason for this omission, it is with pleasure that I welcome and introduce this book by Dr W. D. Foster. It is important that we should have a good account of the period during which the subject passed through successive important phases.

Bacteriology was at first a matter of great public interest, with dramatic new 'discoveries' almost daily – and a chaotic residue to be cleaned up and reorganized before the true was separated from the false. Then it became, in the western world at least, almost a research hobby – full of interest for those who liked it, but not a thing that ordinary doctors greatly needed. Finally, when antibacterial chemotherapy became a reality, the subject was not extinguished – as some providers of finance naïvely hoped and expected – but became an essential discipline of modern medical practice.

Dr Foster takes us to the beginning of this period; and I am grateful to him for putting our record in perspective up to this important point of time – just as the Second World War further increased the importance of bacteriology and before the full development of virology opened up a new chapter.

Dr Foster is well fitted to appreciate and explain to us all these important events and their significance. Apart from the devoted scholarship which has gone into his writing, he brings a distinguished record as a University teacher in Uganda. In the tropical world, every doctor knows that bacteriology must be understood as well as respected.

SIR JAMES HOWIE, LL.D, M.D., F.R.C.P., F.R.C.Path.
Director of the Public Health Laboratory Service
President of the British Medical Association
Past President of the Royal College of Pathologists

Preface

There have been three histories of bacteriology published in English. The earliest was *The Microbe Hunters* by Paul de Kruif. First published in February 1926, a second printing was called for in the March of the same year and the book has since been repeatedly reprinted. This must surely be one of the most successful, if not the most successful of books on medical history. Although written in a popular, journalistic style fully comprehensible to the layman it deserves both its success and the respect of the serious medical historian for it is a pioneer work based on extensive original research. In a second volume, *Men Against Death*, published in 1932, de Kruif added essays on the work of a number of other bacteriologists so that his two books cover in a most interesting way the exciting early history of bacteriology. I will always have the greatest affection for these works which, first read at the age of 15, were instrumental in leading me to take up bacteriology as a career.

However, accurate as they are, de Kruif's books cannot be regarded as a serious history of bacteriology, nor were they intended to be. They remained, however, for several years the only work on the subject, until the completion, in 1931, of the Medical Research Council's nine-volume *System of Bacteriology*. This book contained introductory paragraphs and essays on the history of bacteriology by William Bulloch which were everything the strictest historian could ask in learning and scholarship. In 1936 Bulloch gave the Heath Clarke lectures in the University of London, taking the history of bacteriology as his subject. These lectures were published in book form in 1938 as *The History of Bacteriology* and immediately became the authoritative standard work. It quite soon went out of print and for many years fetched a very high price on the second-hand

market. It was reprinted by the Oxford University Press in 1960.

The third history of bacteriology in English is the small volume written by William Ford for the 'Clio Medica' series which was published in 1939. This volume too has recently been reprinted as a paper-back. Ford's work is written from a point of view slightly different from Bulloch's being more strictly orientated to medical bacteriology and contains, for example, an excellent chapter reviewing the life and work of Koch.

The recent reprinting of the books by Bulloch and Ford suggests that there is an interest in the history of bacteriology and the fact that nothing new has appeared since their publication that it might be useful to bring these histories more up to date.

My book differs from Bulloch's history in two main respects; it is more medically orientated and gives some account of the history of bacteriology from about the year 1900 (where Bulloch ends) to 1938. Bulloch chose to deal with the subject from a very broad point of view. He discussed in some detail the ancient doctrines on the nature of contagion, the development of the idea of a contagium animatum, the microbiology of fermentation and putrefaction and the extensive work, spread over several centuries, to prove or refute the possibility of spontaneous generation of living organisms. All these things are indeed relevant to the history of medical bacteriology but, from the medical point of view, are perhaps dealt with at excessive length, to the exclusion of other interesting material. Thus Bulloch gives hardly any details about the discovery of the important pathogenic bacteria of man, of how they were shown to be causally related to disease and of the use made of these discoveries in the diagnosis, treatment and prevention of disease. As has been already mentioned Bulloch's history ends at about the beginning of the present century when medical bacteriology was about forty years old. My book extends this period by some thirty-eight years bringing the story up to the eve of the era of antibiotics. This period it is true comes after the 'golden era' of bacteriology but was one in which much interesting work was done, particularly on the application of bacteriology to practical medicine. The introduction of

antibiotics, from 1940 onwards, has changed the face of bacteriology so that 1938 makes a convenient point at which to close this history. Moreover there is already available a considerable literature, at all levels, on the history of antibiotics. For example, J. Jaramillo-Arango has a most interesting chapter in his book *The British Contribution to Medicine*, published by E. and S. Livingstone Ltd. in 1953, on the early history of our knowledge of the antagonism which sometimes exists between different species of microbes and shows that this goes back to Pasteur. Recently Professor Ronald Hare, who was working at St Mary's Hospital at the time when Fleming discovered penicillin has, in his book *The Birth of Penicillin*, given a most interesting account of the history of that discovery based not only on researches into the literature and discussion with people still living who worked with Fleming but, also, upon the results of his own repetition of Fleming's observations under various conditions. This work has led, for the first time, to an accurate understanding of this important chapter in the history of bacteriology.

I am most grateful to Sir James Howie for finding time to read my typescript and for consenting to write a foreword for this book. I am also indebted to Professor A. W. Downie, F.R.S. and to Professor R. Hare who have both read and commented on the typescript. I am also grateful to Messrs H. K. Lewis for permission to reproduce plates 1, 6, 7, 9, 10, 12, 16, from Crookshank's *Textbook of Bacteriology* (1896) and also Messrs Constable for plates 13, 14, 15 from Wright's *Technique of the Teat and Capillary Glass Tube* (1912). I should also like to thank Mrs Mary Quick and Miss Ann Ord for typing the book and Mr W. Rivers for preparing the illustrations.

W.D.F.

King's College Hospital Medical School
March 1970

1 The Development of the Germ Theory of Infectious Diseases

Medical bacteriology forms a part of the science of medical microbiology which is concerned with the study of the causative agents of infectious diseases, how the human body reacts towards these parasites and the exploitation of the knowledge gained in the diagnosis, treatment and prevention of this group of diseases.

The parasites causing infectious diseases range from large organisms, clearly animal in nature, such as tapeworms down to ultramicroscopic viruses on the border-line between living and inanimate material and can be grouped into five classes; the helminths, the protozoa, the fungi, the bacteria and the viruses. The last four classes, which are all microscopic organisms, have much in common in the way they produce disease and the way in which the body reacts to them.

The degree of knowledge of these different groups of parasites has, of course, varied independently in time. The larger helminths have been known to be associated with certain diseases of man since ancient times but knowledge of the other groups of parasites has been acquired largely since about 1830. 1840 is the date from which we can begin the study of the history of medical bacteriology in some detail but it is first necessary rapidly to survey the development of our knowledge of infectious diseases and parasites up to that date.

By the beginning of the Christian era many parasites of man and animals had been described and that severe and disabling condition dracontiasis was known to be associated with parasitic guinea-worms. The plague of 'firey serpents' which afflicted the Israelities is thought to have been an outbreak of this disease. But even in such dramatic diseases there was no conception of

the invasion of the body by a parasite from outside (the view most widely held was that parasites developed out of body tissues spontaneously), nor was the parasite necessarily regarded as the cause of the disease (more often it was regarded as the result).

There was speculation as to the part parasites might play in diseases where no actual parasite had been demonstrated, and, for example, sudden death and toothache were sometimes ascribed to 'heart worms' and 'tooth worms'. Nor was the presence of parasites in the body necessarily considered harmful. In times when infestation was very common harbouring some parasites was regarded as normal or even beneficial.

The parasite aetiology of disease received powerful support during the seventeenth century with the invention of the microscope. A whole new world of invisible, living micro-organisms was discovered and such creatures could be better fitted with speculations about the causation of disease than the larger parasites. The German Jesuit Kircher (1602–80), who probably used a simple microscope, and, in a book published in 1658, speculated that plague was due to the invasion of the body by microparasites, was an early enthusiast but his work cannot be considered a scientific advance.

None the less it was during the seventeenth century that the first infectious disease of man was clearly shown to be due to invasion by a microparasitic – scabies. Scabies is caused by the burrowing into the skin of a tiny mite, scarcely visible to the naked eye but easily identified with a simple low-power microscope. This mite was probably known in Europe in the Middle Ages and to the Chinese and Arab physicians but it was not regarded as the cause of the disease and indeed, its existence was regarded as doubtful. The real credit for discovering the mite of scabies and clearly pointing out its aetiological role belongs to the Italian Bonomo, a pupil of the famous Redi. In 1687 he published a very good account which, ten years later, was translated into English and published in the Philosophical Transactions of the Royal Society, and thus given wide publicity. Bonomo's paper is worth considering in some detail. He writes: 'Having frequently observed that the poor children troubled with the itch do, with a point of a pin, pull out of the scabby skin little bladders of water and crack them like fleas

upon their nails, and that the scabby slaves in the Bagnio at Leghorn do often practise this mutual kindness upon one another. . . . It came to my mind to examine what these bladders might really be. I quickly found an itchy person, and asking him where he felt the greatest and most acute itching, I took out a very small white globule scarcely discernible. Observing this with a microscope I found it to be a very minute living creature, in shape, resembling a tortoise, of whitish colour, a little dark upon the back, with some thin and long hairs, of nimble motion, with six feet, a sharp head and two little horns at the end of the snout. . . . With great earnestness I examined whether or no these animalcules laid eggs, and after many inquiries, at last, by good fortune, while I was drawing the figure of one of them by a microscope, from behind a part, I saw a drop, a very small and scarcely visible white egg, almost transparent and oblong, like the seed of a pineapple. . . . From this discovery it may be no difficult matter to give a more rational account of the itch, it being very probable that this contagious disease owes its origin neither to the melancholy humour of Galen, nor the corrosive acid of Silvius nor the particular ferment of Van Helmont, nor the irritating salts in the salts in the serum or lymph of the moderns, but to no other than the continued biting of these animalcules in the skin. . . . From hence we come to understand how the itch proves to be a distemper so very catching, since these creatures by simple contact can easily pass from one body to another, their motion being wonderfully swift, and they as well crawling upon the surface of the body, as under the cuticular, being very apt to stick to everything that touches them, and the very few of them being once lodged, they multiply apace by the eggs which they lay. Neither is it any wonder if this infection be propagated by means of sheets, towels, handkerchiefs, gloves, etc., used by itchy persons. It being easy enough for some of these creatures to be lodged in such things as those, and indeed I have observed that they will live out of the body two or three days.'

Bonomo then went on to point out that it was clear why medicine taken internally was of no value in cases of the itch, and the only remedies that were effective were external applications of ointments made of sulphur, various salts, baths and so on. One can hardly find in the whole of medicine a better

example of a case where knowledge of aetiology pointed more clearly the way to rational means of prevention and treatment of diseases.[1] This discovery of Bonomo's has been claimed to be a turning point in the history of medicine, that it turned doctors from aetiologies based on disturbed humours and made them think in terms of objective, exogenous pathogenic agents as the cause of diseases. Perhaps it should have done but, in my opinion, the wider significance of Bonomo's beautiful discovery was not appreciated and it was certainly another 150 years before another microparasite was shown to cause disease in man.

Two more important discoveries in medical microbiology were made in the seventeenth century; the first observation of parasitic protozoa and bacteria by the Dutch amateur microscopist Anthony van Leeuwenhoeck. Leeuwenhoeck had already, in the year 1676, probably seen creatures which, by their size and from his descriptions, were almost certainly bacteria in infusions of pepper water, but it was not until 1683 that he first described parasitic bacteria in material which he scraped from between his own teeth. Examining this under the microscope he wrote: 'I then again and again saw that there were many small living animalcules in the said matter which moved very prettily.' He then went on to give a description of four different sorts of bacteria, and drew recognizable figures of them. In the 1680s Leeuwenhoeck also described parasitic protozoa for the first time. He found motile animalcules in the gut of a horse-fly and, in his own faeces, 'animalculeae' moving very prettily, some of them a bit bigger, others a bit less than a blood globule. These parasites were in all probability Giardia, and this was, therefore, the first parasitic protozoon to be observed in man.[2]

Belief in parasites as a cause of disease reached a high water mark about 1700. Hartsoeker wrote about that time 'I believe that the worms cause most of the diseases which attack mankind'.[3] Nicholas Andry, a Frenchman, published a most important book *On the generation of worms in the human body* in 1699. This was an exhaustive study of the parasites of man, the diseases associated with them and their treatment. Andry's views were often in advance of his time and, in particular, he did not, like most of his contemporaries, believe in the spontaneous generation of parasites but clearly stated that their

seeds entered the body from without and that some articles of diet were particularly liable to contain them.

New, enlarged editions of Andry's book were published in 1718 and 1741 but, in general, although there was some advance in knowledge of the larger parasites of man and animals, belief in microparasites as a cause of disease waned during the eighteenth century. This was partly due to the imperfections of the microscopes available but probably more to the general conception of the nature of disease and its causation current at the time. The eighteenth century was an age of theories which always went well beyond the observed facts and the notion of a specific cause of a disease necessitates prior conception of specifically distinct diseases. I do not wish to digress into the theories about disease and their classification current in the eighteenth century, but the concept of specifically distinct diseases was lacking, notwithstanding the fact that a beginning had been made in separating the various common fevers, particularly by Thomas Sydenham in the seventeenth century.

The idea that disease consisted of a number of distinct entities, each with its own characteristic clinical picture and characteristic structural change in the organs of the body only began to gain ground in the early nineteenth century and resulted from a new thoroughness of observation of the patient during life and examination of the body after death, with correlation of these two aspects of the disease. It required the birth of the modern spirit of morbid anatomy to lay the foundations upon which specific causation of disease could be based. So important was this new way of looking at diseases, a way bitterly opposed by some of the most eminent doctors of the early nineteenth century, that no history of medical bacteriology would be complete without a digression to consider it in some detail.

The real founder of the doctrine of specificity as we understand it today was Pierre Bretonneau (1778–1862) who practised most of his long life in the French provincial town of Tours. He was a highly original worker with wide interests outside his profession, particularly in natural history. His interest in the specificity of disease is said to have been aroused by observations on the characteristic type of blisters produced by the secretions of different members of the Cantharides group of beetles. He

obtained a junior medical qualification in 1799 and for the next fourteen years practised happily in the country at Chemonaeaux. He acquired such a reputation that he was offered the post of chief physician in Tours and this necessitated him going back to Paris to acquire his M.D. Bretonneau published very little during his lifetime and his reputation, which was very great, depended entirely on the loyal championship of his devoted pupils, particularly Velpeau and Trousseau. Bretonneau's grasp of the concept of specificity derived from his detailed study of two diseases, diphtheria (a name coined by Bretonneau) and typhoid fever. By careful clinical study and performing numerous autopsies Bretonneau showed that the various sorts of diphtheria, whether localized to the throat or nose or spreading to the lower respiratory tract giving the clinical picture of croup, were all one and the same pathologically and quite distinct from the various other inflammatory and ulcerative conditions of the throat. Similarly he separated typhoid from the undifferentiated mass of fevers by pointing out the characteristic lesions in the small intestine. Bretonneau recognized both diphtheria and typhoid to be contagious quoting many clear examples from his extensive experience but he was well aware that neither was as contagious as, for example, smallpox. Trousseau said of his master that 'with a rare order of conception and a sort of intuitive genius Bretonneau wished to apply to the whole of pathology that which he had discovered concerning diphtheria and fever; he worked to prove the specific nature of all diseases. . . . Bretonneau believed that each morbid seed caused a special disease, as every seed in natural history gives rise to a determined species. . . .'[4]

The microscopes available up till the early part of the nineteenth century were inadequate for the effective study of unicellular micro-organisms and until improved microscopes of an essentially modern pattern were introduced in the 1830s no progress in medical bacteriology was possible. The first fatal disease to be shown to be caused by a micro-organism was a disease, not of man, but of a much humbler creature, the silkworm. Silkworms suffer from a number of fatal diseases which can be disastrous to those engaged in sericulture. Amongst these is the disease, which has various local names, but is perhaps best known under the name of muscardine. This disease was investi-

gated by an Italian Agostino Bassi. Bassi was born in Northern Italy in 1773. He was a lawyer by profession, but also managed a small family country estate, and took an active interest in various aspects of farming. He wrote a treatise on warts of sheep and another on the cultivation of potatoes. In the early 1830s he investigated the cause of n uscardine in silkworms and showed quite conclusively that the dead silkworms were infested throughout their tissues with a fungus, and he showed, by innoculation experiments, into healthy silkworms that this fungus was the cause of the disease. Bassi published a book on the disease of silk-worms in 1835, but from that time on, owing to the development of blindness, he did no other microscopical work. However, he had a very clear idea of the possible part that micro-organisms might play in the causation of infectious disease, and he went on to develop this theory as it applied not only to silkworms but to man. He was, for example, an advocate of the isolation of cases of cholera and of the disinfection of the faeces of cholera patients, and of clothing and any other material that they happened to contaminate, and he also advocated the sterilization of the needle used for vaccination, between patients.[5]

The first disease of man that was shown to be associated with a microparasite was ringworm. In 1839 Johan Schoenlein, then 46 years old and professor of medicine at Zürich, described in a paper of only twenty-three short lines, the presence of a fungus in the pustules of a case of ringworm. It is interesting that he was led to this discovery directly by the work of Bassi on silkworms whose work he had taken the trouble to confirm.[6] But he could not be said to have proved that a fungus was the cause of ringworm.

By 1840 the theory that human infectious disease might be caused by microparasites was to some extent 'in the air' but it is, in fact, incorrect to use the term 'infectious diseases' in referring to medicine at that time for this, to us, natural group of diseases, was not, at that time, clearly delineated. Physicians were familiar with diseases which everyday experience showed to be clearly contagious such as smallpox or syphilis; with other epidemic diseases which seemed to be spread by the air such as malaria and yet other diseases such as typhoid fever about the infectious quality of which there was great doubt.

In 1840 an important paper was published by the 31-year-old Jacob Henle, like Schoenlein, of Zürich, entitled 'On miasmata and contagia'.[7] In this long paper which is not remarkable for its clarity of expression Henle reviews all that was known about the origin of infectious diseases but first devotes a great deal of space to establishing the group 'infectious diseases' drawing together and demonstrating the essential unity of the 'miasmatic', 'miasmatic-contagious' and 'contagious' diseases. He insisted on the specific differences between different diseases in this group and considered that it must therefore follow that there were specific differences between their causes. He then adduced evidence that the causes of infectious diseases were living agents which stood 'in the relation of a parasitic organism to the diseased body'. He pointed out that the infectious agent can clearly multiply in the body and pointed to the analogy with ferments which also reproduce themselves and which had recently been shown to be 'lower fungi'. The physico-chemical behaviour of contagious matter, its destruction by heat and disinfectants, also suggested its animate nature. If it was admitted that contagious agents are living creatures they must be assigned to a known group in the biological world and, having dismissed insects as a possibility, wrote that as 'we are daily becoming more acquainted with the wide distribution, the rapid multiplication and the vital tenacity of the lower microscopic plant world, it is even more natural to imagine the contagion as having a vegetable body'.

When he came to discuss the evidence for a particular microorganism as the cause of a particular disease Henle showed a cautious and critical outlook. He accepted Bassi's fungus as the cause of muscardine and the Sarcoptes as the cause of scabies but no other cases as proved. The criteria he demanded as proof of a causal role can be seen from his discussion of the possible role of vibrios (a kind of bacteria) as the cause of syphilis, an undoubtedly contagious disease. He pointed out that the vibrios in question, although very commonly found in syphilitic lesions, were not invariably present and, moreover, similar vibrios had been found in non-syphilitic balanitis and for that matter in material stagnating between the teeth. He would not accept his colleague Schoenlein's paper on ringworm already alluded to because he could not be sure whether the fungus was the

causative agent or merely an organism which found the pustular lesions a particularly suitable place to grow. This approach to the proof of the germ theory Henle communicated to his pupil Robert Koch and was, forty years later, to become one of the corner-stones of classical bacteriology under the name of 'Koch's postulates'.

During the 1840s bacteria and fungi were found in sputum, discharging ears, stomach contents, etc., of man in association with various diseases and the theory that fungi were the cause of many infectious diseases received wide credence. But the critical assessment of such claims advocated by Henle was ignored. Indeed, epoch-making though Henle's paper may seem to the historian looking back at it, I feel some doubt as to the degree of influence it had on his contemporaries. With the wisdom of hindsight it is not difficult to extract from this long paper by Henle, which is not remarkable for its clarity, arguments in favour of a contagium animatum but his contemporaries might be excused for being unable to separate the wheat from the chaff. Henle himself never added any new facts to support this theory and his life's work lay in fields quite other than infectious diseases. His paper seems to have been relatively soon forgotten for an active worker and protagonist of the germ theory, J. B. Sanderson, only came across it by chance in the year 1875.[8]

In 1842 the Edinburgh anatomist, John Goodsir, made a discovery which now seems unimportant but which at the time was regarded as strongly supporting the germ theory. He examined microscopically the vomit of a patient, with clinical features now recognizable as those of pyloric obstruction, and found a highly distinctive micro-organism which he named *Sarcina ventriculi*. Goodsir considered the organisms to be 'either the cause of the symptoms in my patient's case, or at least as very remarkable and important concomitants'.[9] Others soon reported similar cases and therapy designed to inhibit the growth of microbes was said to be beneficial.[10]

The 1840s was a period when the science of chemistry was making rapid advances and particularly through the influence of workers such as Liebig and Andral was being extensively applied to medical problems. Liebig confidently proposed 'a chemical theory of contagion and miasm' and considered that the gravest objection to his idea was 'its simplicity'. He inveighed

against the germ theory as propounded by Henle saying that 'there is nothing more deficient of scientific basis, or more mischievous, than the hypothesis which regards miasms and contagions as animated beings . . .' but his only real reason for his objection seems to have been the sound one that contemporary pathologists were too apt to consider 'two things which occur frequently in conjunction, as standing in the mutual relation of cause and effect'.[11] There indeed seems to have been grounds for this charge; Goodsir's observations on Sorcina ventriculi might be taken as an example.

John Simon in his lectures on general pathology, given at St Thomas's hospital in 1850, discussed Henle's theory which he found 'the utmost difficulty in accepting'. He pointed out that many parasitic diseases of man, animals and even plants were known but they were all essentially different from the contagious fevers whose cause the germ theory sought to explain. True parasitic diseases caused mainly local lesions without generalized body disturbances. The local lesions might indeed be multiple as in muscardine of silkworms but even in that disease, the silkworm did not die until the fungus occupied virtually the whole body. He summarized his objections as follows: 'symptoms are absent, which parasites – if injurious – would unfailingly produce; symptoms are present, which parasites however injurious, could not produce; and, thirdly, the parasites themselves elude discovery'. Simon regarded 'the phenomena of infective diseases to be essentially chemical'.[12] Moreover the state of biological thought at the time permitted alternative explanations, other than a causal role, to be given when the association between a particular microbe and a disease was not denied. Thus Gruby had described the presence of the fungus *Candida albicans* in the lesions of thrush and correctly ascribed to it a causal role. But a critic, who did not deny the association of the fungus with the ulcerated areas in the disease, suggested that 'whenever organic matter or cells, previously endowed with a special form of life, are passing into a state of so-called decomposition, a certain amount yield up their vitality to the overwhelming laws of chemistry and physics, whilst other cells preserve their great endowment of the spirit of life, and take on a new form of organic existence . . . low types of animal and vegetable existence'.[13]

Ernst Hallier, professor of botany at Jena, became a particularly enthusiastic proponent of fungi as a cause of such diseases as cholera, scarlet fever, measles, and typhoid fever, and claimed to have isolated causative fungi. His work was soon shown to be technically unsound and the generally extravagant claims of others mycologically inclined led to the whole germ theory falling into disrepute.

By the 1880s bacteria had been proved to be the causative agent in a number of important diseases of man and his domestic animals and it is easy for the historian to outline the steps leading to the discovery of these relationships in particular cases. What is much more difficult, yet historically of equal importance, is to convey the confusion, complexity and conflicting evidence of the preceding thirty years during which time the 'germ theory' was always to some degree 'in the air'. None the less the attempt must be made.

We may begin by considering the views given on the aetiology of infectious diseases in a classic work of military medicine. A. J. Woodward in his *Outline of the Chief Camp Diseases of the United States Armies as Observed during the Present War*, published in 1863, gives a very interesting account of the infectious diseases seen in the United States Armies at the beginning of the Civil War. Although written by one who had made no special study of bacteria in relation to disease it is worth consideration for the light it throws on the views of the medical profession in general on the aetiology of these diseases. It should be noted that Woodward was well versed in the laboratory aspects of medicine, was a pioneer morbid anatomist and histologist and his work abounds with references to microscopy and chemistry in relation to the aetiology and diagnosis of disease. He is, therefore, unlikely to have overlooked facts in favour of the germ theory which he indeed does discuss. His work, therefore, gives us valuable insight into the position of the germ theory at that time. Woodward deals with the class of disease then known as 'Zymotic diseases' and accepts as a working definition that of Farr 'diseases which are either epidemic, endemic or communicable induced by some specific body or by the warmth or by the bad quality of the food'. The last clause indicates the confused state of thought. Woodward criticized the use of the term 'zymotic' because he believed diseases were in no way

analogous to fermentation which he accepted as caused by micro-organisms. He divided zymotic diseases into the miasmatic, the enthetic (by which he means diseases communicable by innoculation), and the dietetic diseases, but rejects the parasitic diseases which he considered to have nothing in common with true zymotic diseases. In the group of miasmatic diseases Woodward included virtually all of what we would now call infectious diseases but did not consider that they had specific causes, but that they were rather the result of a combination of climate and change in mode of life. In discussing the aetiology of malaria he noted but rejected the cryptogamic theory of its origin and commented that the germ theory had, for some, served 'to explain the nature of epidemic diseases of every kind' but he himself was unimpressed by the evidence. 'Crowd poisoning' was chiefly characterized by the appearance of typhoid, typhus and plague which Woodward considered to be due to eminations given off during respiration, effluvia from the skin and decomposing excreta and diminished atmospheric oxygen. He believed that, according to the prevailing conditions, mixed types of disease occur and appears to have been responsible for the confusing concept of typho-malarial fever. Although he was familiar with Budd's reports on the contagiousness of typhoid fever, Woodward utterly rejected them, saying 'that typhoid presented no phenomena to justify a belief in the possibility of contagion'. One of the most striking epidemic diseases affecting the United States Army during the Civil War was measles and its contagious nature was such as to be undeniable. Regarding measles Woodward said 'it is necessary to acknowledge that although readily spread by contagion when once established the disease may take its origin *de novo* under conditions which are not well understood'. He then went on to review the exactly contemporary work of J. H. Salisbury of Newark, Ohio, who had claimed to have demonstrated that measles was due to a fungus which developed in moist straw. Woodward had himself tried to verify these claims experimentally without success.

The germ theory of the middle of the nineteenth century referred to the possibility that fungi and bacteria were the causative agents. There was naturally no conception of virus diseases in the modern sense. A source of confusion, therefore,

existed in the fact that some of the most strikingly contagious diseases, such as smallpox and measles, which are caused by viruses, could not be satisfactorily demonstrated to be due to bacteria. Examples such as these tended to discredit the whole germ theory. Another difficulty was that it could be shown, by animal experiment, that some bacteria, usually derived from putrifying material, were harmless. To most workers it was not conceivable that such minute and apparently similar organisms could contain a range of distinct species with different properties. Even where an association between microbes and disease was undeniable opponents of the germ theory could not be convinced that the germs had not arisen in the body, as a result of the disease, and were not in any way causally related. This aspect of the theory impinged on the whole problem of the spontaneous generation of micro-organisms, a question which although we today may regard as having been quite settled by Pasteur in the 1860s, if not by Spallanzani almost a hundred years before, did not seem so at the time.

There was also confusion as to whether or not microbes occurred normally in the healthy body. They could certainly be demonstrated in abscesses which were unconnected with the exterior. How did they get there if not by spontaneous generation? Even today there are instances when we cannot give a satisfactory explanation.

These were some of the difficulties which were only gradually removed in certain instances in the 1860s and 1870s. The accumulation of evidence in favour of the germ theory was of two sorts; the proof that more and more diseases were contagious and transmissible by 'something' derived from the sick patient and the definite association of micro-organisms with certain diseases.

Bigelow in 1859[14] divided epidemic diseases into two groups; those clearly contagious such as smallpox and measles and those not contagious such as typhoid and cholera. Epidemics of these last two diseases he thought analogous to illness following the poisoning of a well – an explosive outbreak, but without case to case transmission. Despite the classic demonstration by John Snow that cholera was transmitted by contaminated water and the work of William Budd who tirelessly accumulated evidence showing clearly that the infectious agent of typhoid

fever existed in the faeces of infected patients and was transmitted to the healthy by contamination of their water or food,
the old ideas died hard. None the less Budd's great book *Typhoid
Fever: Its Nature, Mode of Spreading and Prevention* published in
1873 established once and for all the infectious nature of the
disease. Another most important discovery was the proof of
the infectiousness of tuberculosis of J-A. Villemin, a French
army doctor. The occurrence of multiple cases of tuberculosis
in a family was, of course, a common observation but was
universally attributed to a constitutional predisposition to the
disease. Villemin inoculated material from cases of human
tuberculosis into rabbits, showed that they developed the characteristic lesions of tubercle and that the disease so produced
could be transmitted to other rabbits.[15] A young London
physician, W. Marcet, rapidly confirmed Villemin's work and
showed that inoculation of rabbits with sputum could be used
as a diagnostic test for tuberculosis enabling it to be distinguished with certainty for other chest infections. But, by the
medical world at large, Villemin's work was neglected.[16]

The earliest generalized disease of man to be shown to be
caused by a microparasite was trichiniasis. *Trichinella spiralis*
had first been observed by a London medical student, James
Paget, in the muscle of a dissecting-room subject in 1835. The
same parasite had been seen in pork meat in 1848 and by the
beginning of 1860 through the labours of Herbst, Virchow and
Leuckart in Germany the life cycle had been worked out. But
the parasite was in some ways in the position of a modern
'orphan virus' – it was not associated with any disease. However,
in that year, Zenker reported finding numerous fresh Trichinellae
in the body of a girl who had died of an acute illness somewhat
resembling typhoid fever. Several other persons who had eaten
of the same pork were similarly affected but survived. Once the
clinical picture was established numerous outbreaks were
recorded from Germany and the diagnosis proved at autopsy
and by muscle biopsy during life. Here then was a disease of a
systemic kind, quite unlike the local ringworms which had
already been associated with a fungus, which, although not
caused by a germ in the sense of a bacterium, was caused by a
microparasite.[17]

The next major contribution to the germ theory of disease

was Lister's work on the prevention of wound sepsis. Lister had long pondered on the dramatic difference in prognosis between a simple fracture and a compound fracture – one in which the broken ends of the bone had penetrated the skin. The former healed readily but the latter inevitably became septic and often fatal pyaemia followed. Exposure of the broken bones to the air was the main difference and as Lister wrote, in his original paper in the *Lancet* for 1867, regarding 'the question how the atmosphere produces decomposition of organic substances, we find that a flood of light has been thrown upon this most important subject by the philosophic researches of M. Pasteur who has demonstrated by thoroughly convincing evidence . . . (that it is due to) . . . the germs of various low forms of life . . . formerly regarded as merely accidental concomitants of putrescence, but now shown by Pasteur to be its essential cause. . . .' Lister showed quite convincingly that if steps were taken to destroy germs in an open wound with disinfectants and the germs then kept out by antiseptic dressings sepsis did not occur.[18] Lister, although for those early days, a technically expert bacteriologist (he was the first man to develop a technique for obtaining a pure culture) did not trouble much with the details of the bacteriology of wound infection, but his work was striking evidence of the general validity of the germ theory. It should be noted that Lister was not the only surgeon to grasp the significance of Pasteur's work on fermentation and putrefaction in relation to surgical sepsis. Another British surgeon, T. Spencer Wells, in an address at the British Medical Association meeting in Cambridge in 1864, three years before Lister's first publication on the antiseptic system, drew attention to the work of Pasteur, to Davaine's work on anthrax and to the various microparasites such as Trichinella, Schistosoma and the ringworm fungi which were already known to be associated with human disease and suggested that micro-organisms in the air might be the cause of surgical sepsis. Unlike Lister, however, Wells did nothing to prove his hypothesis correct.[19]

One of the best-known contributors to the germ theory controversy in the middle of the nineteenth century was L. S. Beale who held, in succession, the chairs of physiology, pathology and medicine at King's College Hospital medical school. Beale, who was born in 1828, was a brilliant exponent of the use of the

microscope in medicine. Indeed his chief claim to be remembered is as a pioneer clinical pathologist. His textbook *The Microscope in Medicine*, first published in 1854 and which ran to four ever-expanding editions by 1878 is one of the classics of clinical pathology. In 1868 Beale was invited to give a course of lectures in Oxford and chose to speak about 'Disease germs, their origin and nature'. His discussion was based upon his very extensive practical experience of the microscopic examination of human and animal tissues. Beale believed that 'The higher life is, I think, everywhere inter-penetrated as it were by the lowest life. Probably there is not a tissue in which these germs do not exist, nor is the blood of man free from them. . . . So long as the higher living matter lives and grows, the vegetable germs are passive and dormant, but when changes occur and the normal condition departs, they become active and multiply.' He correctly showed that epithelial cells of the digestive and respiratory tract might indeed contain bacteria and a combination of post-mortem artefacts and inadequate technique convinced him that all cells were similar in this respect. In objecting to the germ theory Beale pointed first to the ubiquity of bacteria and fungi which suggested their harmlessness. It was not for another ten years that Koch proved that pathogenic bacteria were quite distinct from saprophytic forms and the concept that there might be specific differences amongst bacteria morphologically similar seems to have been grasped by very few. And the fact that undoubtedly infectious matter, such as material from a fresh smallpox vesicle, did not contain bacteria and, moreover, tended to lose its infectivity if bacteria developed in it was against the idea that bacteria caused disease.

As an alternative to the germ theory Beale suggested, bringing forward much personal experimental work as well as speculative arguments, that the disease germs were derived from the living 'bioplasm' of the body which, instead of producing normal tissues was perverted to produce 'disease germs'. Contact between healthy tissues and a 'disease germ' led to the former producing more 'disease germs'. The closeness of Beale's theory to the modern view of the nature of virus infection will be obvious.[20]

The first bacterium causing a disease in man to be described was the leprosy bacillus. This was reported by the Norwegian,

Armaur Hansen, in 1874. Leprosy was at that time common in Norway and that country was the European centre for the study of the disease. As well as local experts, distinguished pathologists from other parts of the world often went there to investigate the disease. Leprosy was generally regarded as hereditary and not contagious. It is Hansen's chief merit that he insisted, in the face of opposition, on the contagious nature of the disease. He undoubtedly saw the leprosy bacilli but his techniques were crude and he was in doubt as to the bacillery nature of the objects he saw and of their causal role. It was one of Hansen's visitors, Neisser, the discoverer of the gonnococcus, who first gave a clear account of the organism based on a study of properly stained material, in 1879.[21]

The study of one particular disease, anthrax, probably did more than anything, once and for all, to establish bacteria as a cause of disease and its history is worth considering in some detail. Anthrax is a relatively unimportant disease of man but an important disease of sheep and cattle. The first report of bacteria in the blood of a sheep dead of anthrax was made by a Paris physician P. F. U. Rayer in 1850, the actual observations having been made by his friend C. Davaine. The same bacteria were independently discovered by a German, P. A. A. Pollinder, and the infectivity of anthrax blood amply demonstrated by several workers. Davaine's interest in anthrax, which had lain dormant since 1850, was rekindled by the reading of Pasteur's work on butyric fermentation with its description of the rod-like bacterial bodies responsible for this reaction. Davaine was struck by their similarity to the organisms he had seen in anthrax blood. Taking up the subject again, in a series of more than twenty papers, between 1863 and 1870, he aroused general interest in the subject and established the bacilli as the cause of the disease. He by no means, however, solved all the problems of the epidemiology of anthrax. Davaine, a charming, modest man, was one of the foremost parasitologists, in the broadest sense of the word, of the nineteenth century. Apart from his work on anthrax he wrote a comprehensive textbook on the parasites of man and animals and studied the parasitic diseases of various plants.[22]

The further study of the anthrax bacillus introduces us to one of the greatest bacteriologists of all time – Robert Koch. Koch

had qualified at Gottingen University where he had been the pupil of J. Henle and so was probably familiar with his master's views about bacteria in relation to infectious disease. He had not elected to take up an academic career and was, at the time of his anthrax studies, in his early thirties and a general practitioner in a village in eastern Germany. Koch was familiar with Davaine's work but still regarded the relationship of the bacteria to disease as somewhat doubtful and particularly felt that the bacilli could not survive long outside the animal body. Since the contagium of anthrax certainly maintained itself over winter in certain fields Koch reasoned that the bacilli must form spores, organs which were already known in saprophytic bacteria. His work was primarily designed to demonstrate this phase of the life cycle. He inoculated mice with anthrax blood and maintained the infection through twenty generations from the original material. He cultivated the anthrax bacilli in beef serum or aqueous humour of an ox-eye, on a microscope slide in a home-made moist chamber and incubator. He arranged his apparatus so that he could continuously watch the development of a particular bacillus and saw the development of highly refractile spores which he likened to a string of pearls. Taking a preparation of spores dried on a coverslip he added a drop of aqueous humour and watched the development of bacilli from spores and yet a further generation of spores from these bacilli and he showed that the inoculation of dried spores produced anthrax in mice. Koch showed further that exactly similar bacilli, found in hay, which also produced spores did not cause anthrax on inoculation this demonstrating clearly, for the first time, a species difference with respect to pathogenicity of bacteria. In Koch's own words, 'It follows, therefore, that only a species of bacillus is able to cause this specific disease, while other schizophytes have no effects or cause entirely different diseases when inoculated. . . .'[23] Koch demonstrated his experiments, at the University of Breslau in April 1876, to F. Cohn, actually professor of botany but one of the founders of bacteriology. Cohn did all he could to forward Koch's interests and published Koch's work in his own journal the same year.

Despite the clear proofs of the causal relationship of the bacillus of anthrax to the disease, brought forward by Davaine and Koch, there appears to have still been some doubts on the

matter. Transmissions had always been made with the blood of an infected animal to a healthy animal and some still argued that it was not the bacilli but something else in the blood which caused the disease. Even Koch's careful work had not fully satisfied this objection and to do so was left to one who must be regarded as the greatest of all microbiologists – Louis Pasteur. Pasteur grew anthrax bacilli for many generations in purely artificial culture, in sterile urine, and showed that, at the end, the bacilli still produced anthrax on inoculation into a guinea-pig.

During April and May 1875 the Pathological Society of London devoted several sessions to a discussion on the germ theory of disease and a report of these meetings occupies some ninety pages of the society's Transactions. This is a particularly interesting report since from it may be grasped the position of the germ theory less than three years before its general validity had been established. The main contributors were Charlton Bastian, the 38-year-old professor of pathology at University College Hospital and J. B. Sanderson, his 37-year-old colleague the professor of physiology. There were a number of minor speakers two of whom deserve mention at the outset. An elderly veterinarian named Crisp quoted the remark of Cullen that 'there are more false facts than false theories; and probably those who hereafter read this discussion will say that it is a confirmation of the statement I have made . . .' This contribution seems to have been unappreciated but exactly describes the feeling of the historian nearly a hundred years later. The second minor contributor deserving of notice is F. Payne, physician and pathologist at St Thomas's Hospital. He confessed that he knew little about bacteria but raised the possibility that there might be specific differences between pathogenic bacteria and those associated with putrefaction. Speaking as a morbid anatomist, he observed, that in pyaemic abscesses, in which bacteria could be shown to be present, there was never any malodorous gas which was so characteristic of the putrefactive process.

The main speaker against the germ theory, C. Bastian, immediately set about illustrating the truth of Mr Crisp's comment. He admitted the association of bacteria with certain disease processes but not 'to the extent alleged'. He doubted the

validity of the analogy between infectious disease and fermentation and was sceptical about Pasteur's claims that fermentation was due to living micro-organisms. Bastian considered that it was 'generally admitted' that (a) bacteria could be introduced into experimental animals without ill-effect, (b) bacteria were present in healthy living tissue, (c) lesions produced in experimental animals by chemical irritants were full of bacteria, (d) there were contagious fluids which undoubtedly did not contain bacteria and the development of bacteria in such fluids reduced the virulence of the material, (e) bacteria were not present in the blood during life but soon appeared there after death. Bastian maintained therefore, that even the very existence of organisms in the fluids and tissues of diseased persons 'is for the most part referable to the fact that certain changes have taken place (by deviation of healthy nutrition) in the constitution and vitality of such fluids and tissues, and that bacteria and allied organisms have appeared therein as pathological products . . .' The most life-like manifestation of a contagious fluid was its obvious multiplication in the body of a new host but Bastian suggested that this was, not the multiplication of a living organism, but a process analagous to the growth of crystals in a strong salt solution; the addition of a minute crystal fragment initiated the process of crystallization which then continued on its own. It was also possible that one living cell could change into another by a process of 'unfolding of organic forms'.

Sanderson, although he had spent many years in the experimental study of bacteria in relation to disease and was probably Britain's foremost exponent of the germ theory, made a very lame reply to these apparently cogent arguments. He would have preferred not to discuss such things as the origin of germs and confined himself to 'questions of disease, not questions of biology'. He by no means disposed of Bastian's arguments. He quoted the particularly good example of micro-organism showing a very close association with the manifestations of disease; the spirillum of relapsing fever. This organism had been discovered in the blood of relapsing fever patients by the young Berlin physician Otto Obermeier in 1873. His observations had been confirmed and it had been shown that the spirilla were present in the blood during a febrile paroxysm, i.e. when the

patient was ill but not in the intervening periods when he was afebrile.

Eventually the discussion was wound up at a late hour and it is probably true, in so far as the sense of the meeting can be taken from the printed record, that opinion was against rather than for the germ theory. Even two years later a leading article in the *Lancet* reviewed the whole question of the germ theory without coming to any firm conclusions.

2   The Contribution of Louis Pasteur to Medical Bacteriology

The science of medical bacteriology, during the latter half of the nineteenth century, was dominated by the work of two great men and their immediate pupils – Louis Pasteur and Robert Koch. Although Pasteur was over twenty years older than Koch and had accomplished great things in the broad field of microbiology before Koch ever commenced his researches, it happens that their most important contributions to medical bacteriology were both made at about the same time, during the decade from 1876 to 1886. Both were profoundly interested in the validity of the germ theory in general and made important contributions to this topic, but, by and large, their work hardly overlapped. Koch's work lay chiefly in the identification of bacteria as a cause of human disease and Pasteur's chiefly in the field of immunity. When they began work about 1876 medical bacteriology did not exist, yet ten years later it was an established branch of medicine. So much did Pasteur and Koch each achieve that the most convenient way of dealing with this particular period in the history of bacteriology is to devote separate chapters to the contributions of these two great bacteriologists. We will deal first with the work of Pasteur.

Pasteur was born in 1822 at Dole. His father was a tanner and ex-soldier of the Napoleonic army. He was, by education, a chemist and held teaching posts at Dijon, Strassburg and Lille before settling at the Ecole Normale in Paris. His first research was in the field of crystallography and, as is well known, a connecting thread is clearly visible from this work, done in the 1840s, to his last great achievement; vaccination against rabies. The nature of racemic acid led to an interest in fermentation and the discovery of the part specific micro-organisms played in lactic and butyric fermentation. This, in turn, led to

his investigations into the cause of spoilage in wines and beer – the 'disease of wine'. From the diseases of wine, Pasteur turned to his successful studies on the diseases of silkworms so that, by the late 1870s, no man was better prepared for the study of the causes of the infectious diseases of the higher animals and man.

In 1877 Pasteur was already 55 years old and had behind him almost thirty years' study of microbes in their various aspects other than as causes of disease in man. He was a firm believer in the germ theory of disease and had supported Davaine in his opinions about anthrax in 1863. Moreover, in 1867, Lister had developed his antiseptic system acknowledging that it was based on the fundamental researches of Pasteur into the cause of putrefaction. Pasteur's interest had thus gradually been drawn towards the aetiology of infectious diseases. His earliest contribution to human pathology seems to have been an offshoot from the controversy over spontaneous generation. Pasteur had been able gradually to demolish the observations of those who adhered to the old theory but the clear demonstration of bacteria in pus from abscesses, which had no connection with the exterior of the body, indeed seemed an unequivocal example of spontaneous generation. Pasteur argued fiercely before the Academy of Medicine that this just could not be so but was not able to offer a satisfactory alternative explanation.

Pasteur commenced his work on infectious diseases with at least three advantages, excluding genius. He had unshakeable faith in the germ theory so that, despite apparently contradictory evidence, whether brought forward by opponents or arising from his own observations, he never wavered. His work with beer and wines had accustomed him to making pure artificial cultures of micro-organisms which we have already seen contributed to the establishment of bacteria as the cause of anthrax. Lastly he had thoroughly grasped the concept of specific differences between micro-organisms and was accustomed to note small morphological differences and distinct physiological properties. His work during the ten-year period under consideration can be broken down into three slightly overlapping periods; the works he accomplished can be listed and each dealt with in more detail subsequently. Between 1877 and 1879 he investigated septicaemia due to Clostridium septicum and

the epidemiology of anthrax; discovered the causative organism of boils and osteomyelitis as well as that of puerperal fever, and engaged in the general defence of the germ theory before the academy of medicine. Between 1879 and 1881 he was largely occupied working on the attenuation of the microbes causing chicken cholera and anthrax and developing effective prophylactic vaccines. During this period he also contributed to the study of plague, pleuropneumonia, cholera and rouget des porcs. From 1881 to 1886 he was wholly occupied in work on the experimental transmission of rabies and the development of vaccines against that disease.

Clostridium septicum infection

Early in 1878 Pasteur was engaged in acrimonious discussion at the Academy of Medicine on the germ theory and its application to surgery and, in the April of that year, gave a lecture to the Academy of Science on the subject in which he gave details of his researches into the disease produced by an organism he called 'Vibrion septique' (*Clostridium septicum*).

It had been known for some time that if experimental animals were inoculated with a variety of putrid materials some animals would die and their death be associated with the appearance of bacteria in the blood. Pasteur attempted to cultivate this organism artificially, initially without success. He used sterile urine, yeast water and 'bouillon de viande' – the ancestor of today's universal, nutrient broth. He particularly recommended 'bouillon Liebig'. It then occurred to him that the organism might be a strict anaerobe (he was familiar with anaerobic organisms from his work on butyric fermentation). This supposition proved to be correct the bacterium growing well in a vacuum or under carbon dioxide. Exposure to air killed the organisms. He observed the formation of spores (with which he was again familiar from his work on flascherie – a disease of silkworms) and realized that, in this condition, the microbes were resistant and could be blown about by the wind and survive in water. He performed inoculation experiments which clearly showed that the microbe was the cause of the fatal septicaemia and that it was different in a number of respects from the bacillus of anthrax. Here was then yet another example of a disease due to a bacterium.[24]

The epidemiology of anthrax

Pasteur's demonstration that a pure culture of the anthrax bacillus in an artificial medium would cause anthrax in an experimental animal had put the finishing touches to the work of Davaine and Koch on the aetiology of that disease. Pasteur was able, by studies in the field, to work out the epidemiology of the disease. At the request of the Minister of Agriculture he spent the summer of 1878 in the department of Eure-et-Loire trying to work out how sheep caught anthrax under natural conditions. He tried to transmit the disease by feeding sheep with buzeme deliberately contaminated with anthrax bacilli but was hardly ever successful. But if the food was of a rough character, containing, for example, thistles which caused wounds in the mouth, such feeding stuff, if contaminated with bacilli, gave rise to anthrax. A study of the morbid anatomy of spontaneously occurring anthrax also suggested that the disease began in the region of the oro-pharynx. He showed that anthrax bacilli could survive a long time on the ground and that they could be recovered from the surface of the ground ten months after an animal had been buried in that area, even after deep burial. In one experiment four sheep were kept over the grave of an animal dead of anthrax and buried over two years before. One of the sheep died of anthrax on the eighth day and Pasteur was able to show, by guinea-pig inoculation, the presence of anthrax bacilli in the surface soil and particularly in the casts of earthworms, it being these creatures which transport the anthrax spores from a carcass deeply buried. He suggested as a preventive measure that animals dead of anthrax should never be buried in a field that was to be used for pasture or that they should be buried in dry silicous or calcarious soils which were unsuitable for earthworms.[25]

The aetiology of puerperal fever

In March 1879 a clinician named Hervieux read a paper at the Academy of Medicine on the cause of puerperal fever in which he denied that bacteria, which were found throughout nature and 'quite unoffensive', could cause disease. Pasteur rose to say that he felt obliged to reply in defence of the germ theory and went over much of his old work. He then recorded

that, in 1875, when already familiar with the presence of cocci in abscesses, he saw a case of pyaemia following a septic abortion in which he found cocci in pus from an abscess during life and also from the blood after death. The idea had then occurred to him that puerperal fever might also be caused by such cocci.

The day after this discussion Hervieux allowed Pasteur to examine material from a severe case of puerperal fever in his wards. He found, in the lochia, round organisms occurring in pairs or chains just as he had postulated at the meeting. He cultured the same organism from the blood of the patient both during life and after death. He was able to confirm his observations on another case and also to show that the lochia of healthy women did not contain these organisms. Pasteur concluded, therefore, that the organisms he had found were probably the cause of puerperal fever. One complicating factor against his thesis was that, in addition to these organisms, he had seen other motile rod-shaped organisms. Hervieux maintained that he could only believe in germs as a cause of puerperal fever if a single specific microbe could be implicated. Pasteur fell back on his 'faith' in the germ theory for it must be admitted that his reply was not satisfactory. He maintained that with the great diversity of clinical picture in puerperal fever it was not surprising that more than one organism might be involved. None the less there can be no doubt that Pasteur was correct and that he had seen and duly implicated the Lancefield group A streptococcus, the most important cause of puerperal fever.[26]

The discovery of Staphylococcus

In May 1879 Emile Duclaux, one of Pasteur's colleagues, was suffering from a severe crop of boils. Pasteur had one of them punctured (he did not care to do this sort of thing himself) and the pus obtained was inoculated into meat broth and incubated at 35° C. Next day the broth was turbid and under the microscope he saw an organism which he recognized as distinct from the others he knew. It was round, occurred in pairs, rarely in fours but often in little masses. This organism was undoubtedly the *Staphylococcus aureus*. Pasteur was able to confirm his original observation on more of Duclaux's boils and on other patients. He also showed that despite the widespread nature of the furunculosis the blood remained sterile. He

regarded this organism as the cause of boils. A few months later he was allowed to examine some fresh pus from a case of osteomyelitis when he again isolated a coccoid organism. Limited though the characters of the organism available for study were, Pasteur recognized this as the same organism as he had isolated from boils and confidently announced that osteomyelitis was, in effect, 'a furuncle of the bone marrow'![27] Pasteur is not usually credited with the discovery of the Staphylococcus, the honour going to a 37-year-old Aberdeen surgeon, Alexander Ogston. This is probably just since Pasteur made no further observations on this subject and Ogston's entirely independent observations, published in 1881, were much fuller from the clinical, bacteriological, pathological and experimental point of view. Ogston named the organism staphylococcus because its arrangement was reminiscent of bunches of grapes. Three years later a German surgeon, Rosenbach, distinguished the three species of staphylococcus, *S. aureus, S. albus* and *S. citreus,* on the basis of the pigment produced in artificial culture on solid medium.

The production of vaccines against chicken cholera and anthrax
The years 1879 and 1881 Pasteur largely devoted to the work of attenuating the virulence of the causative organisms of chicken cholera and anthrax and demonstrating that effective prophylactic vaccines could be prepared. Of all his wonderful contributions to science this was certainly the most useful and, in many ways, the most original and yet it seems to have been 'dashed off' in a very short time with hardly a hesitation or setback – the result of happy chance, the 'prepared mind' and genius. It is well worth while looking at this aspect of his life's work in detail.

Pasteur was interested in any new example of disease proven to be due to a microbe, as material to consolidate the germ theory as a whole. It was therefore natural that when Perroncito, Professor of Pathology at Turin, announced that an epidemic disease of fowls commonly called 'chicken cholera' was due to a bacterium that Pasteur should be interested. Pasteur was, as he himself confessed, less interested in such details as to whether an organism was a coccus or a bacillus but vastly concerned with what different microbes could do and how they interacted

with their environments, be it a fermenting wine or the blood-stream of an animal. He had already pondered on some observations on the difference in susceptibility of different animal species to anthrax, particularly the immunity of fowls to this disease, which he had shown depended upon their higher body temperature. He had demonstrated that lowering the body temperature of a fowl rendered it susceptible to anthrax. Probably in the early spring of 1879 Pasteur, during a few days rest in the country, had read with attention the works of Edward Jenner on vaccination against smallpox published over eighty years previously. This was before his own experiments had led him into the field of artificial immunity. This preamble is an attempt to account for Pasteur's 'prepared mind' which chance was soon, dramatically, to favour.

The microbe which causes chicken cholera, which we now call *Pasteurella septica*, had been first seen about 1874 but first adequately described by Perroncito in 1878. In the same year Perroncito's work was confirmed by a young Toulouse veterinarian, H. Toussaint, who had gone further and managed to cultivate the organism using the medium introduced by Pasteur for anthrax, neutralized urine. Pasteur, in repeating this work, found that neutral urine was not a very satisfactory culture medium but that a broth made of chicken meat was excellent. He also noted that the chicken cholera microbe would not grow in yeast infusion as did the anthrax bacillus. He likened this difference to differences in natural immunity between species and, from a practical point of view, used failure to grow in yeast infusion as a test to distinguish the Pasteurella from saprophytic organisms. He also noted that guinea pigs were relatively immune to Pastueurella infection. Individual chickens also showed some variations in resistance for even his most virulent cultures sometimes failed to kill 100 per cent of inoculated animals.[28]

We have the authority of Emile Duclax, one of Pasteur's colleagues at the time, for the accidental circumstances which led to the discovery of a means of attenuating the virulent Pastuerella. It was simply that their work was interrupted by the vacations. Cultures which had been put on one side for some weeks were found no longer to be capable of killing chickens. It then occurred to Pasteur to reinoculate these

chickens with a fresh young culture. In Duclaux's words 'to the surprise of all, perhaps even of Pasteur himself, who did not expect such a success, almost all these chickens resisted, whereas new chickens, just brought from the market, succumbed in the ordinary length of time. . . . What secret instinct, which spirit of divination impelled Pasteur to knock at this door, which was only waiting to be opened? Here we see clearly the part played by his readings and his former studies, by the incessant ponderings which had been going on in his mind, and by the intervention, in the midst of these obscurities, of this faculty of imagination. . . .' Pasteur immediately saw the analogy between his observations and the old practice of variolation against smallpox, the inoculation of cattle with pleuro-pneumonia material and with Jenner's work, using cow-pox material, as a preventive inoculation against smallpox. But he was quick to observe a new principle. In all these old vaccination procedures the infecting agent, loosely spoken of as a virus, was unknown. Chicken cholera was caused by a known bacterium which could be cultivated artificially. Bacterial diseases were therefore fundamentally the same as the 'virus diseases' and there was no reason why the immunization procedure, so successful in preventing smallpox, should not be extended to all known bacterial diseases. But he appreciated the specificity of the immunity, for he showed that chicken cholera immune fowls could still be infected with anthrax.

The veterinarian, Toussaint, seems to have been quick to see the possibilities of immunization against other bacterial diseases and, in 1880, reported successful immunization of cattle against anthrax using heated and filtered anthrax blood as vaccine. It was natural, therefore, for Pasteur to attempt to produce immunity to anthrax. His initial opportunity to demonstrate that such immunity did exist came in the summer of 1880, when he was asked to assess a form of treatment, proposed by a certain veterinarian, for anthrax. Cattle were deliberately infected with anthrax by Pasteur and although no greater proportion of the treated cattle survived than the controls, Pasteur thus obtained a small supply of cattle which had survived an attack of anthrax. He showed that a second inoculation of virulent anthrax bacilli did not even make the cows sick. It is interesting that at this point Pasteur recalled an experience, in

1878, for which his mind, at that time, was evidently not prepared. He had noted that some sheep which he had infected
with anthrax-contaminated food had subsequently been shown
to survive large inoculations of anthrax blood. He could thus
enunciate a rule for anthrax, as for chicken cholera, that
inoculations of organisms, which did not kill, protected against
subsequent infection. But, at that time, September 1880, he
did not know of any way to produce a culture of anthrax
bacilli which could be relied upon not to kill. The problem,
however, was one which he solved in a remarkably rapid
manner.

Mere ageing of the culture attenuated the chicken cholera
microbe but this method was inapplicable to anthrax, old cultures of which merely produced resistant spores. It seemed to
Pasteur that he must discover some methods of getting an aged
culture of anthrax which did not produce spores. Such a problem one might suppose would take no little time to solve yet,
by February 1881, Pasteur had devised a method and shown
that such aged, sporeless cultures were indeed avirulent. After
a few preliminary experients with antiseptics, Pasteur found
that simply growing anthrax bacilli at 42° C., a few degrees
higher than the optimum temperature, prevented spores formation and indeed, after a month under such conditions, the
bacilli were quite dead. However, after eight days growth they
were not dead but were avirulent and in fact all degrees of
virulence could be obtained by adjusting the period of culture.
It seems probable that the germ of the idea, of growing the
organism at a raised temperature, was planted in his mind by
the work of Toussaint alluded to earlier.

Pasteur had scarcely time to run a preliminary experiment in
the laboratory when the president of the Melun Agricultural
Society suggested a decisive, public experiment to test the
efficacy of his preventive inoculations, his society providing the
necessary animals. Pasteur accepted confidently. 'What succeeded with fourteen sheep in the laboratory will succeed with
fifty in Melun.' The story of the experiment at Pouilly-le-Fort
must be one of the best known tales in bacteriology and little
imagination is required to picture the dramatic situation. A
group of animals was vaccinated twice, with a twelve-day interval, with avirulent anthrax cultures. Fourteen days later the

vaccinated animals and an equal number of unvaccinated controls were inoculated with virulent anthrax bacilli. The results of the experiment were judged two days later on 2 June 1881. Pasteur may be excused some momentary anxiety and regret at his audacity but the feeling did not last long. As Roux, one of his collaborators in this experiment, wrote, 'The next day, more assured than ever, Pasteur went to verify the brilliant success which he had predicted. In the multitude which thronged that day at Pouilly-le-Fort, there were no longer any who were incredulous; only admirers.'[29] The results may be quoted directly from his report to the Academy of Science on 13 June. 'The 24 sheep and the goat which had received the attenuated virus as well as the 6 cows had all the appearances of health; on the contrary, 21 sheep and the goat which had not been vaccinated were already dead of anthrax; 2 other unvaccinated sheep died before the eyes of the spectators and the last of the series died at the end of the day.'[30]

In August 1881 Pasteur was invited to the International Congress of Medicine, held in London, where he gave a short address explaining in simple language his work on the attenuation of chicken cholera and anthrax microbes. He expressed himself well satisfied with the progress of the germ theory in England, indeed of its 'triumph'. Referring to his principles of vaccination it was, he said, 'a method the fruitfulness of which inspires me with boundless anticipations. . . . May we not be here in presence of a general law applicable to all kinds of virus? What benefits may not be the result? We may hope to discover in this way the vaccine of all virulent diseases. . . .' Pasteur ended with a graceful compliment to the memory of Edward Jenner, saying 'I have given to vaccination an extension which science, I hope, will accept as a homage paid to the merit and to the immense services rendered by one of the greatest men of England, Jenner.' It should be noted that the word 'vaccination' coined by Jenner was etymologically appropriate for inoculation of cow-pox material into man, being derived from the latin word for a cow – vacca. It was not particularly appropriate to the procedures introduced by Pasteur but it was part of his homage to Jenner to establish the term which is now used with reference to prophylactic immunization of all kinds.[31]

The economic importance of anthrax led to immediate wide-spread use of Pasteur's prophylactic inoculations which, in general, were highly successful. But it was not to be expected that in the hands of the multitude the experience of Pouilly-le-Fort would be everywhere repeated. Acrimonious criticism came from the great German bacteriologist, Robert Koch; criticism of a personal nature and sometimes justified in detail. But history can only remember it to Koch's discredit that he failed to appreciate the wonderful principle established by Pasteur, in what was little more than a few months work. It is not too much to say that the principle enunciated by Pasteur, at that London Congress in 1881, was the greatest single discovery in the history of medicine, comparable with, if not more important than, Lister's discovery of the principle of antisepsis in surgery.

Remarkable though the work described was it must not be thought that it constituted the whole of Pasteur's contribution to medical microbiology during the years 1879 to 1881. He made contributions to the study of plague, pleuropneumonia and cholera and discovered the causative organism of rouget des porcs. But here we must remember that Pasteur had attracted to himself devoted and able colleagues whose help he himself always acknowledged. It is certain that without their help his achievement would have been nothing like as great. Four of his assistants at this period must in justice be named, Emile Duclaux, later Director of the Pasteur Institute, a chemist and administrator whose book *Pasteur – the History of a Mind* is the best published study of Pasteur's work; Emile Roux, another director of the Pasteur Institute, Foreign member of the Royal Society; Copely, medalist and, after Pasteur, perhaps the greatest French bacteriologist; Charles Chamberland whose name we still remember in association with a type of bacterial filter and Louis Thuillier, who having collaborated with Pasteur in the discovery of the cause of rouget des porcs, died at the age of 27 in Egypt, of the cholera he had been sent to investigate.

Vaccination against rabies
We must now consider the work which, more than any of his previous researches, established Pasteur's popular fame – his discovery of a preventive inoculation against rabies. It is not

quite clear why Pasteur chose to work on this problem and yet in some ways it was particularly well chosen. Rabies is not an important disease of man. The annual mortality from it is trivial and even this can be abolished by simple quarantine measures. Also, since when man is infected, the mortality is invariably 100 per cent, there was no means of knowing that immunity against the disease was even possible. Further, mass immunization against such a rare disease would never be possible and only the fact that rabies had an exceptionally long incubation period allowed the hope that immunization, after being bitten by a rabid animal, might be of some value. For all these reasons the choice of rabies as the first human disease against which to develop a vaccine seems curious. From the point of view of furthering the cause of prophylactic immunization rabies had the advantage of seeming, in the eyes of the general public, to be a dramatic disease. But rabies was a considerable veterinary problem and perhaps Pasteur, who lacked any qualification in human medicine, felt more at ease with a disease that was on the border-line between human and animal medicine. Also, in 1879, Galtier, a veterinarian at Lyons, had shown that rabies could be transmitted to rabbits by inoculation of the saliva of a rabid dog.[32]

Pasteur's first experiment in the field of rabies seems to have been done on 11 December 1880 when he inoculated rabbits with mucus taken from the mouth of a child dead of the disease. The rabbit died within thirty-six hours but of a disease which Pasteur did not for a moment confuse with rabies. The disease of which the rabbit died was a septicaemia and its blood contained, in abundance, a new type of organism. Under the microscope it appeared as a short, rod-shaped organism, somewhat narrowed in the middle and surrounded with a halo of gelatinous material. The organisms could be cultivated in various sorts of meat broth but then lost its halo and formed long chains. This organism was almost certainly the pneumococcus and this account was the first published description of the organism, although the American, G. M. Sternberg, had produced exactly the same results in rabbits, with an inoculation of his own saliva, three months earlier. Pasteur seems to have spent the first three months of 1881 studying his new type of septicaemia on which he published several more papers.[33]

Pasteur first reported the successful transmission of rabies to rabbits in May 1881, and went further than Galtier in that he successfully demonstrated that the virus existed in the spinal cord as well as the saliva. During the next eighteen months he added a number of new facts to knowledge of the disease. He maintained the virus by serial passage in rabbits in the laboratory; developed the technique of intracerebral inoculation which, unlike subcutaneous inoculation, invariably caused infection; demonstrated that human brain tissue from rabies cases also contained the virus and took his first step towards developing a vaccine.

He found that one out of three experimentally inoculated dogs did not die and that they subsequently resisted even introcerebral inoculation of virulent material. He showed that virus could be found in the nerves, blood and in the salivary glands and also that it survived for weeks in spinal cord tissue outside the body provided it was kept cold to prevent putrefaction. Despite the apparently infectious nature of the virus material he was quite unable to cultivate a microbe and, moreover, does not seem to have wasted time chasing any contaminating bacteria. During the year 1883 Pasteur made a most important discovery; he found a way of altering the virulence of the rabies virus. The virus taken from the disease as it occurred naturally in dogs took about fifteen days to kill a rabbit, after intracerebral inoculation, but, by repeated passage through rabbits, its virulence was increased until, eventually, it would kill in eight days, but no less. Pasteur termed the original virus 'street virus' and the passaged virus 'fixed virus'. Further, passage through one species of animal enhanced the virulence for that species but not for others – he could produce at will viruses of different virulence for different species of experimental animal. In particular he found that virus passaged through a series of monkeys became less and less virulent for dogs and rabbits until it was harmless. Dogs inoculated with such virus subsequently proved resistant to virulent virus. Likewise, dogs inoculated with material of gradually increasing virulence for rabbits became immune, even to intracerebral inoculation of virulent virus. Pasteur suggested that it might be possible, in human medicine, to profit by the long incubation period of the natural disease and build up an artificial immunity.

However his method of protecting dogs was not always successful, immunity being achieved in only about three-quarters of the animals. Searching for a better method Pasteur hit upon one of the utmost simplicity. He found that if rabbit spinal cord, containing 'fixed' virus, was suspended in dry air over caustic potash the virulence of the material gradually abated and a series of cords of decreasing virulence could be obtained. If dogs were inoculated with the least virulent first and daily given cord material of increasing virulence they became completely immune to virulent virus. Using this method he immunized fifty dogs without a single failure. His method was ready for trial in a human patient.

On 6 July 1885 Joseph Meister, aged 9, who had been bitten by an undoubtedly rabid dog on 4 July, was brought to Pasteur. He was severely bitten and, in consultation with his doctor and a professor of medicine, it was agreed that his death was inevitable. Pasteur was therefore fully justified in trying his prophylactic measure. Joseph Meister was given a subcutaneous injection of cord material which had been dried fifteen days. During the following ten days he received injections of cord material dried for shorter and shorter periods. Meister survived, not only the natural rabies infection acquired from his bites, but, by the end of his course of injections was receiving virus which was actually more virulent than 'street' virus.

Pasteur reported the result in a paper to the Academy of Science on 26 October 1885 and, this time at least, not a single critical comment was heard. Pasteur had chosen a disease where 'one swallow' did 'make a summer'; not one person, bitten as Joseph Meister had been bitten, had ever survived before. By the beginning of March 1886 Pasteur was able to report the results of 350 cases treated by his method. Of course there was often doubt that the animal inflicting the bites in a particular case really was rabid, despite the most stringent inquiries. Of these 350 patients only one died. But although some of the patients were probably bitten by animals not, in fact, rabid, there was no reasonable doubt as to the efficacy of Pasteur's preventive measure, for two reasons. Firstly at the end of their vaccination course the patients were actually being inoculated with fully virulent virus and, secondly, over a number of years, it had been shown that about one in six persons

bitten by supposedly rabid dogs did die of rabies. Formal statistical tests were hardly necessary. Pasteur ended his paper by saying that a preventive against rabies following bites had been demonstrated and that a vaccine establishment for this purpose should be founded.[34]

The scientific interest and importance of Pasteur's work on rabies was only surpassed by its popular appeal (despite the relative unimportance of the disease), an appeal probably not equalled even by medically far more important discoveries such as that of penicillin. Funds for Pasteur's vaccine establishment were immediately forthcoming from all over the world and, by November 1888, the President of France opened the fine, new, well-endowed Pasteur Institute on the Rue Dutot. From the beginning the interests of the Institute were far wider than the prevention of rabies. It immediately became the premier microbiological institute in the world for teaching and research and, despite many other foundations since, probably remains so.

Pasteur himself lived on until 1895, his last year dogged with much ill-health. But in tracing the history of his unsurpassed contributions to medical microbiology we do not need to look further than 14 July 1885, the day Joseph Meister survived an inoculation of fully virulent rabies virus.

3 The Contribution of Robert Koch to Medical Bacteriology

Robert Koch was born in the Harz mountains in Germany, in 1843, son of a mining engineer. As a boy he was a keen naturalist and for the whole of his life took an amateur interest in botany. He studied at Göttingen University where he had a thorough scientific education including student research experience. After qualification he did a variety of jobs, including service in the army, before settling at the age of 29, in the remote village of Wollstein in East Germany. Koch's working life falls into several well-defined periods which we may enumerate and then consider in greater detail.

As soon as he settled in Wollstein as a general practitioner he set aside part of his consulting-room for microscopical work. It has been said that his wife saved up the money to buy his first microscope from her housekeeping money but whether this is true or not I do not know. Certainly it would be a mistake to regard Koch as poor and he seems to have equipped himself with a very good little laboratory. It was during his eight years at Wollstein, in his leisure moments, that Koch laid the foundations of his reputation as a bacteriologist and, indeed, had he never done anything else but his work at Wollstein he would still be regarded as one of the greatest bacteriologists of all time. At Wollstein Koch did the work on anthrax which we have already considered, worked on technical matters, such as staining and photomicrography of bacteria, and published a classic paper on the bacteriology of wound infections. This distinguished work led to an appointment at the Reichsgesundheitamt in Berlin, nominally as a physiologist, in 1880, when he was 37 years old. During the following five years Koch made the discoveries which laid the foundations of clinical bacteriology. He described his technique for obtaining pure cultures on solid media, certainly the most important single contribution ever

made to the science of bacteriology. He made fundamental studies on the action of disinfectants and discovered the causative organisms of tuberculosis and cholera. At the age of 42 he was made professor of hygiene in the University of Berlin and later director of a newly-founded Institute of Infectious Diseases but made no more discoveries comparable with his days at Wollstein or at the Reichsgesundheitamt. He started the first course in practical bacteriology and through his pupils continued to develop his subject. The names of his pupils, like Henry V's Agincourt role, are, to bacteriologists, 'familiar in our mouths as household words', Leoffler who isolated the diphtheria bacillus, Gaffkey who first cultured the typhoid bacillus, Ehrlich the great immunologist and founder of scientific chemotherapy, von Behring the discoverer of antitoxin and von Wassermann whose name must, even today, be mentioned at least once a day in every hospital in the world – all these, and many more, were Koch's pupils and some of the credit for their work belongs to him.

There is an element of sadness in the twenty years of Koch's maturity. He made no more discoveries of the first importance and in the field of tuberculosis, where his original work had been so brilliant, he caused disappointment and confusion by publishing a worthless method of treatment and by denying the importance of cattle as a source of human infection, errors all the harder to overcome because of Koch's great authority. About 1890, at a time when Koch's claims for his tuberculin treatment of tuberculosis were being sharply criticized and he was under much pressure of responsibility and overwork, he got into the habit of making regular visits to a theatre in the neighbourhood of his laboratory. Here he made the acquaintance of a young actress with whom he fell in love. Such was his passion that he obtained a divorce from his first wife and married his actress. This act did Koch great harm in scientific and social circles, particularly as Koch's autocratic behaviour had made him not a few enemies. Metchinkoff recalls that, at the 1892 Congress of medicine, Koch's new marriage formed the main topic of conversation and excited more interest than the scientific papers.[35]

Koch was principal medical adviser to the German Colonial Office and, from 1896 onwards, seems to have managed to spend

a great deal of his time in the tropics investigating mostly diseases caused by protozoa rather than bacteria. In 1904, when aged 61, he resigned his directorship of the Infectious Diseases Institute so as to be able to spend most of his time abroad. From boyhood he had had the ambition to travel to exotic places and study obscure tropical diseases. His later years were probably some of the happiest in his life when, perhaps, he managed to recapture some of the thrills of the old Wollstein days. It is likely that Koch, who was, at heart, a simple character was much happier investigating exotic diseases in the tropical bush with one or two devoted, younger colleagues than playing his Excellency the Director in high political and social circles in Berlin. Tuberculin was a great disappointment to him and he may well have felt, bitterly, that had he been but left alone to continue his researches in peace there would never have been such a humiliating fiasco. It cannot be said that Koch's tropical investigations achieved anything considerable and, indeed, his authoritative pronouncements were sometimes unfortunate. But some interesting observations he did make. He went to British South Africa to investigate rinderpest of cattle and also showed that the 'coast fever' of cattle was caused by a similar protozoon to that which caused Texas Fever. He studied plague in India and malaria in Java, Sumatra and Malaya. The scene of Koch's most extensive tropical experience was East Africa which he visited for prolonged periods on three occasions , in 1897–8, 1905 and 1906–7. He worked mainly in Tanganyika in co-operation with the German medical authorities there. Although no longer young and the amenities of East Africa few he trekked about indefatigably. There was a mysterious outbreak of fever in the Usumbara mountains; Koch arrived with his microscope (the first ever seen in that area) and quickly identified the fever as malaria and drew attention to the greater susceptibility of the inhabitants of the mountains to that disease as compared with the habitual plain-dwellers. He was given a room in the new medical laboratory at Dar-es-Salaam where he discovered that monkeys harboured a kind of malaria parasite and trained a young attendant, named Hassani Selemani, who continued work in the laboratory until his death in 1941. Koch's subsequent work in East Africa was largely concerned with sleeping sickness and, in 1906, he

suggested that trypanosomes actually underwent a cyclical growth phase in tsetse flies and were not merely transmitted mechanically. Unfortunately he went on to describe a sexual cycle analagous to that of the malaria parasite which does not, in fact, occur. Koch studied the distribution of the disease, which appeared to have spread southward from Uganda, and thought that crocodiles were the most important hosts for the tsetse flies. He tested the new arsenical compound, Atoxyl, on sleeping sickness patients which he found effective but was, unfortunately, too optimistic in his statement that it was as good as quinine in malaria. He lived and worked in an abandoned Church Missionary Society station on Bugalla in the Sesse Islands.[36] In August 1906 he visited Entebbe, where the British sleeping sickness commission had its laboratory, and in the following year, came back again this time to Kampala where he stayed the night with the Cooks at Mengo Hospital. Sir Albert Cook records that Koch was 'modest and unassuming', and 'made a thorough inspection of our hospital'.[37] Koch was offered space in the new British sleeping sickness commission laboratory but this had recently been moved, on account of the fears of the inhabitants of Entebbe, to an isolated hilltop at Mpumu – a twenty-three-mile rickshaw drive from Kampala. Koch regretted that he could not accept the offer and privately remarked to the senior medical officer that although the laboratory was magnificent the position was quite impossible, and went on to say that if any German colonial official had put up a laboratory in such a place he would be dismissed the service.[38] Koch was shocked to find that the laboratory in Uganda was not open on Sundays 'but his efforts to have it opened seemed to be regarded as irreligious. He stirred up a theological controversy there among the missionaries at Mengo by propounding his view that it was only after the Fall that idle man had seized on the idea of resting every seventh day, in order to avoid honest labour'. *Nunquam otiosus!!* In Makerere medical school library there is an interesting memento of this visit. Sir Albert Cook took Koch's photograph which he pasted, with a little note, in a book on the tuberculin treatment of tuberculosis.

For some years prior to his death Koch suffered from occasional attacks of some circulatory disturbance but had no thought of retirement. He was at work until the last, reading his

last scientific paper only six weeks before his death from cardiac failure, in May 1910.

The Wollstein period

Koch demonstrated his anthrax work in F. Cohn's botany department at the University of Breslau in April 1876 and his paper was published the same year. In the following year he published the second important contribution of his Wollstein period – a paper on the technique of staining bacteria for microscopical examination and photography. Looked at today the subject seems very elementary but it must be remembered that at the time the most experienced bacteriologists examined their bacteria fresh and unstained, in a drop of culture fluid, and with relatively lower powers of the microscope. We have seen that this technique was employed by Pasteur and we can but marvel that he was none the less able confidently to distinguish streptococci, staphylococci and pneumococci.

It was Koch who introduced the technique, now used universally, of making thin smears of bacteria containing fluid on glass slides or coverslips, fixing and staining them and examining them under high-powered oil-immersion lenses. Today we simply heat our dried films to fix them, but Koch, who was anxious to preserve bacteria as little distorted as possible, tried various fluid fixatives eventually settling on an aqueous solution of potassium acetate. Weigert had already shown that haematoxylin was a suitable stain for bacteria but Koch introduced the use of analine dyes of which such a large range was becoming available. The methyl violet and fuchsin which are used every-day in a modern bacteriological laboratory were found by Koch to be the best. For the purpose of photomicrography he found analine brown gave the best results. Koch took excellent photographs of bacteria and pointed out that they often enabled a closer study to be made than the actual slides. On some of his preparations he could see bacterial flagella. Such stained preparations could be preserved, sent through the post and examined by many people. He even suggested that many of the false observations and assertions 'would not be published to swell the bacteriological literature to such a turbid stream if, for each object seen, the investigator had to submit preparations for demonstration.'[39]

The last piece of Koch's work dating from His Wollstein period was his 'Investigations into the Etiology of traumatic infective diseases'. This beautiful series of experiments and observations, described and discussed with the utmost lucidity, was immediately recognized as the 'classic' it is and published in English in 1880, the year after its publication in Germany. Reading it is a delightful experience which has something of the flavour of Theobald Smith's account of his investigations of Texas fever.[40]

The group of diseases Koch proposed to investigate were the supposedly infective diseases which complicated injuries and surgical operations such as local suppuration, erysipelas, septicaemia and pyaemia. But, with only the material and facilities of Wollstein available to him, Koch abandoned any idea of working with human material and set to work to reproduce analagous diseases in rabbits and mice.

He was by no means the first to study the bacteriology of this group of diseases. Rindfleisch, in 1866, had noted bacteria in metastatic abscesses in a case of pyaemia, as had Birch-Hierschfeld and Klebs in septic wounds. All observers agreed that bacteria were difficult to distinguish microscopically, using the crude techniques available, and Koch was at an advantage having already developed good staining techniques. Other workers had tried to transmit the human infective diseases to animals with varying degrees of success. Confusion often arose because an animal might die of an intoxication from the putrid material rather than in infection. None the less Davaine had produced a clearly defined infectious septicaemia and Orth had transmitted human erysipelas to rabbits. Klebs had gone even further and induced septicaemia with artificial cultures of micrococci from septic infections. Koch, in reviewing all this earlier work, admitted that the evidence in favour of bacteria as the cause of the traumatic infective diseases was very strong but there were a number of objections which might legitimately be raised. Firstly, it was by no means settled whether or not bacteria existed in normal animal tissues. If they did their causal role in infection became very dubious. Koch did not consider this a serious objection being firmly of the belief that, provided technique was satisfactory, blood and tissues would always be found to be sterile. Other objections were better

founded. There was, for example, considerable doubt about the constant association of bacteria with septic disease and even when present, they were found in relatively small numbers. This last fact was in contrast to the one disease which Koch accepted as definitely due to bacteria, anthrax; in this disease the tissues swarmed the bacteria. The third point was that the bacteria found in septic processes were morphologically very diverse and bore striking resemblances to saprophytic organisms. The position here was summed up by Birch-Herschfeld who wrote, 'the morphological characters of the bacteria found in pyaemia, diphtheria, smallpox and cholera are so similar that the idea naturally arises that identical organisms are being dealt with. But, if this were the case, it would follow that no specific significance could be attributed to these forms. They would have to be regarded merely as parasites of the disease and not as its cause.' Koch went on to point out that micrococci indistinguishable from each other had been found in diseases as diverse as erysipelas, puerperal fever, endocarditis, rinderpest and pleuropneumonia. Since it was hardly credible that the same coccus caused all these diseases, either the presence of cocci was incidental or (almost revolutionary thought!) similar-looking cocci might, in fact, be different. What was necessary was 'that the presence of bacteria in these diseases be proved without exception, and further that the conditions as regards their number and distribution be such as to afford a complete explanation of the symptoms' also 'we require conclusive evidence that this or that micrococcus, definite in nature and always recognizable under varying conditions by certain characteristics is the only cause of the disease in question'. Thus was Koch feeling his way towards his famous 'postulates'.

Koch's own experimental work was entirely histological using the essentially modern method of fixation of tissues followed by sectioning and staining with analine dyes. He did no culture work in this piece of research which was actually an exploitation of the improved staining and microscopical technique which he had previously reported. He paid particular attention to the illumination of his slides and appears, after experiments of his own, to have introduced the Abbé condenser to the bacteriological microscope. He also introduced the useful staining principle of overstaining a tissue section and then

decolorizing it with potassium carbonate solution which, he found, removed all the dye from the tissues but not from the bacteria. This method was later to prove of crucial importance in his investigation of the cause of tuberculosis.

Koch obtained his experimental diseases by inoculating a variety of putrid materials into rabbits and mice and obtained six different diseases each caused by its own specific microbe. He produced:

1. *Septicaemia in mice.* This disease, coming on after an incubation period of twenty-four hours, gave a characteristic clinical and pathological picture. It was transmissible from mouse to mouse by the smallest scratch of an infected scalpel. This disease was associated with a minute bacillus in the blood and tissues which, whilst highly virulent for mice, was harmless to rabbits. Koch observed and described the phenomenon of phagocytosis (four years before Metchnikoff) but did not grasp its importance. He thought the bacteria penetrated and multiplied within the leucocytes. He also described the migration of the leucocyte in inflammation without recognizing its significance.

2. *Progressive gangrene in mice.* This condition often occurred at the same time as septicaemia. However Koch showed quite clearly that the diseases were unrelated. The gangrene was associated with a chain-forming coccus which remained near the site of inoculation and did not invade the blood-stream. He watched the pathology of the gangrene in the mouse's ear and postulated that the organisms produced a soluble toxin causing tissue necrosis at some distance from the masses of cocci. At first he was not successful in separating the gangrene-producing cocci from the bacilli of septicaemia, since transmission of the former inevitably involved infection with the latter. However, inoculating a field mouse, he found that whereas the cocci produced the usual gangrenous lesion, the bacilli failed to develop in that species.

3. *Spreading abscess in rabbits.* A disease caused by yet another coccus which he found in great numbers at the advancing edge of the abscess but which never invaded the blood-stream or produced alterations in the internal organs. These cocci were morphologically quite distinct from those causing gangrene in mice, being much larger and growing in masses rather than chains.

4. *Pyaemia in rabbits.* He produced this disease by the inoculation of macerated mouse's skin and the pathological picture was not unlike pyaemia in man. This disease was associated with micrococci and he particularly noted the tendency of the blood to coagulate in the vicinity of the organisms. There were few organisms in the blood as they tended to settle out in the capillaries.

5. *Septicaemia in rabbits.* This disease was also associated with cocci in the blood but was clearly different from the pyaemia from a morbid anatomical point of view. The cocci caused little local reaction, were smaller and showed no tendency to cause coagulation of the blood.

6. *Erysipelas in rabbits.* Koch produced this disease by inoculating the ear with mouse dung. This disease was associated with a bacillus which produced severe local inflammation but showed no tendency to invade the blood-stream.

We must remember that at the time (1879) there was but one disease that had been convincingly shown to be caused by a bacterium – anthrax. Koch had added six more demonstrating that each disease, admittedly experimental, but closely resembling some of those seen in human surgical practice, with its distinct pathological picture was caused by a distinct microbe. He confidently predicted that the analagous diseases of man would be found to have bacterial causes. Koch regarded, and history would agree with him, the most important aspect of his work as the establishment of the 'differences which exist between pathogenic bacteria and to the constancy of their characters. A distinct bacteria form corresponds, as we have seen, to such disease, and this form always remains the same, however often the disease is transmitted from one animal to another.' Koch anticipated that this assertion would be 'much disputed by botanists, to whose special province this subject really belongs'. He quoted one eminent botanist as saying, in 1877, that, 'I have for ten years examined thousands of different forms of bacteria, and I have not yet seen any absolute necessity for dividing them even into two distinct species.' But, in reality, the germ theory in its general sense was proved beyond all doubt by the work described in this monograph by Koch. The research had impressed upon him the vital importance of being able to make pure, artificial cultures of bacteria and of the limitations of the

fluid cultures such as were used by Pasteur. He wrote that to prove that a particular bacterium was the cause of a disease, when present mixed with others, it would be necessary to isolate it in pure culture and show that it produced the disease in question. Koch never laid down as formal rules the 'postulates' which have been attributed to him although, in 1884, in a paper on tuberculosis, but using anthrax as an example, he explained the criteria necessary before it could be assumed that an organism caused a particular disease. It seems that these criteria were forced upon him in thinking about his experimental infections for we have seen them expressed in essence in different parts of that work.

It is impossible to overpraise this work, the product of a general practitioner's leisure moments. It can stand by the side of any paper in the history of microbiology from Pasteur's on prophylactic vaccines to the cracking of the genetic code.

Kaiserliche Gesundheitamt period

When Koch took up his appointment at the Kaiserliche Gesundheitamt in Berlin, in 1880, he was provided with three assistants all of whom proved most apt pupils and became themselves bacteriologists of the first rank. They were only a few years younger than Koch himself. The eldest was G. Gaffky who eventually succeeded Koch as head of the Institute for Infectious diseases, and among many important contributions was the first to cultivate the typhoid bacillus. Next came F. Loeffler, a man whose range of contributions to science few others have equalled. He cultivated the bacilli of diptheria and glanders, recognized the first mammalian virus (that of foot and mouth disease), introduced numerous technical procedures (remembered today, for Loeffler's methyline blue and Loeffler's medium), was an excellent teacher and wrote on the history of bacteriology. Finally there was G. Wolffhugel, who was eventually professor of Hygiene in Göttingen and died when only 45. These three men must always be associated with Koch's work during his Gesundheitamt period.

The responsibility of the health department to some extent dictated or at any rate suggested Koch's next line of research which was to attempt to place the processes of disinfection on a scientific basis. But before considering this, we must look at a

long paper published by Koch, in 1881, entitled 'on the investigation of pathogenic organisms' in which he reviews the experience of his former years of work and lays down the methods by which further work must be done. Part of this technical work we have already discussed and seen the fruits of his approach in his investigations on wound infections. It is only a portion of the 1881 paper which we must now consider, that on *Pure Cultivations* for it embodies the most useful single contribution to bacteriology of all time. Koch regarded the ability to produce pure cultures as absolutely essential if bacteriological knowledge was to advance and yet, writing of the methods then available commented 'on the whole it is truly depressing to attempt pure cultivations' . . . and a practical impossibility to take all the necessary precautions. Without better methods no results could be accepted as convincing. In particular Koch thought his remarks 'especially applicable to the researches (carried on with really remarkable, if blind, zeal) now issued in quantities from Pasteur's laboratory, and which describe incredible facts with regard to pure cultivations of the organisms of hydrophobia, sheep-pox, pleuropneumonia etc.'[41] This last quotation is included illustrating, as it does, an unhappy side of the history of bacteriology in the nineteenth century. It is sad and inexcusable that the two greatest bacteriologists of their day should have had such personal animosity between them. Koch's remark was nothing short of insolent when addressed to a man who had made fundamental advances, involving the use of pure cultures, whilst Koch was still a schoolboy and, who, at the time of Koch's attack, was laying the foundations of the science of immunology. Moreover, other able, contemporary bacteriologists thought that Koch exaggerated the difficulty of making pure cultures in liquid media and, as we have seen, Pasteur had already distinguished the streptococcus of puerperal fever from the staphylococcus of a boil and shown the identity of the latter with the causative organism of osteomyelitis using such cultures.

But to return to Koch's paper on pure cultures; we will give an account of his discovery largely in his own words. He wrote 'it being perfectly clear that efforts in this direction were in vain I have abandoned the principles on which pure cultures have been hitherto conducted and have struck out on a new

path to which I was led by a simple observation which anyone can repeat.

'If a boiled potato is divided and the cut surface is exposed to the air for a few hours and then placed in a moist chamber . . . it will be found by the second or third day . . . on the surface of the potato numerous and very varied droplets almost all of which appear to differ from each other. A few of these droplets are white and porcellanous, while others are yellow, brown, grey or reddish and while some appear like a flattened drop of water others are hemispherical or warty. . . . If a specimen is taken from each of these droplets so long as they remain distinctly isolated from each other and are examined by drying and staining a layer of it on a coverslip it will be seen that each is composed of a perfectly definite kind of micro-organism.

One, for example, will show enormous micro-cocci, another very minute ones, a third might show micro-cocci, arranged in chains, while other colonies, especially those just spread out flat, like a membrane, are composed of bacilli of various size and arrangement.' Koch was not the first to observe, study or publish an account of the pure colonies of different microbes that could be obtained on the cut surface of a potato. J. Schroeter, a worker in Cohn's department, with whose work one might have expected Koch to be familiar had done all these things in 1872, but Koch makes no reference to this work. It was Koch's great merit to appreciate the enormous possibilities of solid media in general and to show that with their aid pure cultures of bacteria could be grown regularly with the greatest of ease. Potato itself proved surprisingly useful, even for pathogenic bacteria, and the early textbooks of bacteriology abound with coloured plates illustrating the growth of various species on this medium. But Koch, realizing its relative limitations, set about, not looking for other solid media, but for a means of solidifying well-known liquid media. Gelatine was an obvious substance to try and mixed with ordinary nutrient broth gave 'nutrient gelatine'. Best of all for pathogenic bacteria was sterile serum solidified with gelatine. Koch fully appreciated the enormous possibilities of solid media for isolating, for counting and for distinguishing different species of bacteria. No longer were there only differences in morphology and physiological effects to observe, but also the nature of the pure colony which he soon

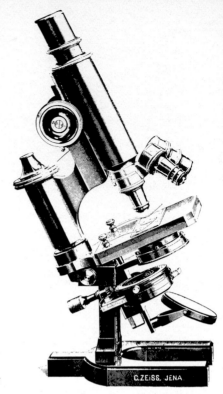

1. A common model bacteriological microscope, circa 1890.

Crookshank, 'Text-book of Bacteriology' 1896

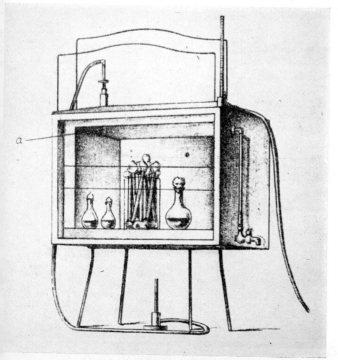

2. A bacteriological
incubator, circa 1885.

*Woodhead and Hare
'Pathological Mycology' 1885*

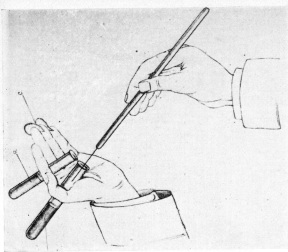

3. Methods of inoculating early bacteriological media.

Woodhead and Hare 'Pathological Mycology' 1885

(a) Tubes of nutrient gelatine

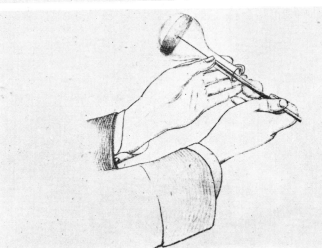

(b) Flasks of bread paste

(c) Slices of potato

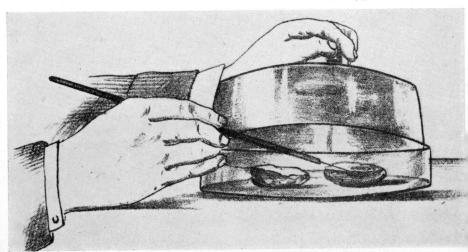

showed varied considerably yet were always constant for a given organism.

Koch tried pouring his nutrient gelatine into various containers for use but eventually settled on the method whereby he smeared the molten nutrient gelatine on sterile glass slides which could then be kept moist under a bell-jar. His method of sowing bacteria on them was to dip a sterilized platinum needle into the culture and then draw the point rapidly several times over the surface of the gelatine 'in much the same way as the lancet in vaccinating by incision . . . in fact this process might very well be called an inoculation'. And so it still is to this day.

Minor contamination of Koch's plates did not matter merely producing one or two isolated colonies which were readily recognized for what they were. The chief disadvantages of nutrient gelatine were that it was liquid at body temperature and that some bacteria digested it. The latter could be a useful distinguishing property but was also a nuisance when dealing with mixed organisms. Koch summed up his claim to the novelty of his method thus: 'The peculiarity of my method is that it supplies a firm and where possible a transparent pabulum and its composition can be varied to any extent and suited to the organism under observation, that all precautions against the possibility of after contamination are rendered superfluous, that subsequent cultivation can be carried out by a large number of single cultures of which, of course, only those cultures which remain pure are employed for further cultivation and that finally a constant control over the state of the culture can be obtained by the use of the microscope. In almost all these points my method differs from those hitherto employed and especially also from the former attempts at the cultivation with potatoes and isinglass referred to above.'

Nutrient gelatine had a short life as a bacteriological medium, the disadvantages alluded to above proved particularly irritating, to none more so than W. Hesse who had studied under Koch and who was, in his home laboratory, studying the various bacteria found in the air. Tired of having his cultures spoiled by gelatine-liquifying organisms he looked for other means of solidifying his culture media. Fortunately his wife, Fannie, as well as her housewife's duties acted as her husband's technician and prepared broth for the bacteria as well as soup

for the family. She suggested the use of agar-agar, an extract of Japanese seaweed, which she had used for some years for culinary purposes. It proved completely successful. Hesse wrote to Koch informing him of this and Koch immediately took it up. No formal paper on the use of agar was ever published, Koch merely mentioning it in a short sentence in his preliminary paper on tuberculosis in 1882.[42] Neither was Koch's technique of smearing molten medium on glass slides very convenient but it was not until 1887 that R. J. Petri, an assistant of Koch, introduced what he called 'a slight modification' of the method and described the covered glass dish now used universally which bears his name. Surprisingly it was some years before Petri's dish ousted Koch's plates. [43]

Chemical disinfectants of various kinds had been used empirically for a long time. They varied from a nosegay of sweet smelling herbs on the bench before an assize judge to the burning of sulphur in the sickroom or the addition of carbolic acid to sewage. But there was no exact knowledge of the germ-killing power of any disinfectant. Nor, until recently, had there been any knowledge of the nature of the contagia which were supposed to be destroyed by the disinfectants. Koch tried to put the whole matter on a scientific basis. He appreciated that disinfectants should kill bacteria rather than inhibit their growth, that different organisms might be expected to differ in susceptibility – in particular, that spores were likely to be more resistant than vegetative organisms, and that the effect of a disinfectant would be influenced by the time it was allowed to act and its concentration. He therefore devised tests that would take account of all these variables.

It was convenient to use as test organisms bacteria which could be readily recognized and distinguished from any contaminants. Koch therefore chose, as representative of vegetative organisms, two strikingly pigmented organisms *Serratia marcescens* and *Pseudomonas aeruginosa* and for a spore bearing organism, the anthrax bacillus. He exposed fragments of potato on which his vegetative organisms were growing to the action of a disinfectant and then applied them to a fresh piece of potato and noted whether or not growth occurred. The anthrax spores were impregnated on to silk thread and after exposure to disinfectants and washing in sterile water planted on nutrient gelatine.

One of the first disinfectants Koch subjected to his tests was carbolic acid, for some fourteen years the basis of Lister's antiseptic system. He found that although 1 per cent aqueous carbolic acid killed vegetative organisms in two minutes and a dilution of 1 in 850 prevented the development of spores it required the action of a 5 per cent solution for two days to actually kill spores. Further, he found that Lister's 5 per cent carbolic acid in oil failed to kill anthrax spores within three months. Because of the relative inefficiency of the commonly employed disinfectants Koch tested a very large number of chemicals. He noted that the presence of organic matter such as blood serum interfered with the action of some disinfectants and concluded that the only really satisfactory substances were chlorine, bromine and mercuric chloride.[44]

Koch and his colleagues also turned their attention to disinfection by heat, noting the relative inefficiency of dry heat as well as the variability in temperature in different parts of ovens and its poor penetrating power. They also studied the sterilizing power of steam both at atmospheric and increased pressure. Finding that even spores were readily killed by moist heat at 100° C. they recommended steam, at atmospheric pressure, rather than the mechanically less satisfactory apparatus for dry heat at higher temperatures or for steam under pressure. This was not particularly clever work, and some of the laboratory findings conflicted with clinical experience, but it laid the foundations of the scientific study of disinfection.

Of the many problems impinging on a public health department such as the Kaiserliche Gesundheitamt none was greater than that of tuberculosis. Statistics showed that about one seventh of human beings in civilized parts of the world died of this single disease. There was no doubt at that time (1880) that tuberculosis was an infectious disease. We have seen that Villeman showed that tuberculosis could be transmitted to animals by inoculation in 1865. This work had been amply confirmed and Tappeiner had successfully transmitted tuberculosis to dogs by inhalation.[45]

Koch therefore decided to seek a bacterial cause for tuberculosis using the staining techniques which had proved so successful in his studies of the traumatic infective diseases. He examined smears of tuberculous material on coverslips as well as

sections of tuberculous tissue. Initially he used tuberculous lesions from infected guinea-pigs of which he could conveniently maintain a supply. Using exactly the same staining techniques as he had used in his previous work, Koch was, at first, unsuccessful in demonstrating any organism in tuberculosis material. But, being certain that such an organism must exist, he persisted with his efforts trying various ways of altering the staining procedure. By what steps he actually came upon his successful method he does not record but part of the method depended on a technique which he had previously discovered; that of overstaining and then decolorizing until only the bacteria remained stained. Koch's technique was to stain his material in methylene blue to which potassium hydroxide had been added; it was the addition of the alkalie which made the tubercle bacilli take up the stain. At first he had to leave his preparations in the stain for twenty-four hours but later found that the staining time could be reduced to one hour if done at 40° C. By this time the whole preparation was deeply stained blue but if it was then rinsed for a few minutes in an aqueous solution of a brown, analine dye called vesuvin, the methyline blue was washed out of everything except the tubercle bacilli which could easily be seen as blue bacilli against a brown background. Koch noted at once that the bacilli which he could see were characteristic in size, shape and arrangement and were unlike any other organism he had worked with and he also noted their close similarity with the bacilli found in leprosy. From experimental tuberculosis in guinea-pigs Koch went on to examine material from all types of tuberculosis in man and animals and always found his characteristic bacillus. On these grounds he felt that it was probably the causative organism.

But, 'in order to prove that tuberculosis is brought about through the penetration of the bacilli, and is a definite parasitic disease brought about by the growth and production of these same bacilli the bacilli must be isolated from the body, and cultured so long in pure culture, that they are freed from any diseased production of the animal organism which may still be adhering to the bacilli. After this the isolated bacilli must bring about the transfer of the disease to other animals, and cause the same disease picture which can be brought about through the

inoculation of healthy animals with naturally developing tubercle material.'

As with their staining, so the tubercle bacilli proved unusually difficult to grow in artificial culture and the many failures which Koch must have experienced, using techniques which were quite satisfactory for other pathogenic organisms, might well have daunted a less patient and persistent worker. Koch devised a quite new medium, cow or sheep's blood serum solidified by heating to 65° C. Even on this medium, whereas all the common organisms with which he was familiar grew within twenty-four hours, no growth of tubercle bacilli was visible for ten days or more. Initially Koch isolated tubercle bacilli from experimentally produced lesions in guinea-pigs and was able to show that the isolated bacilli when inoculated into healthy animals caused tuberculosis, just as did naturally occurring infective material. He then went on to isolated bacilli, with just the same characteristics, from various types of human and animal tuberculosis.[46]

Koch presented his findings at a meeting of the Physiological Society of Berlin (probably because officially he was physiologist to the Gesundheitamt) on 24 March 1882. He concluded his paper by saying:

'All of these facts taken together lead to the conclusion that the bacilli which are present in the tuberculous substances not only accompany the tuberculous process, but are the cause of it. In the bacillus we have, therefore, the actual tubercle virus.'

It is said that Koch's communication was greeted with stony silence – there was none of the lively discussion which commonly follows the presentation of a scientific paper. But, after all, what was there to discuss? Koch had done it all, no possible doubtful point had been left unresolved. Rarely in the history of science can an important discovery have been so incontrovertibly presented, once and for all.

Koch's discovery ushered in a new branch of medicine, that of clinical bacteriology. For it was immediately appreciated that the demonstration of the characteristic bacilli in, for example, sputum, offered a new diagnostic tool of unparalleled precision. Within a few weeks of Koch's announcement of his findings, Ehrlich published a paper describing an improved staining technique and the results he had obtained when examining specimens of sputum. He used analine water and methyl violet or

fuscin instead of alkaline methyline blue as his primary stain and shortened the time of action to fifteen to thirty minutes only. He deliberately decolorized everything except the tubercle bacilli by rinsing the preparation for a few seconds in 30 per cent nitric acid, before counterstaining with a yellow or blue dye. This method stained the organisms more intensely than that of Koch. The bacilli appeared larger and more were shown up. He found the bacilli every time in the sputum of twenty-four patients with tuberculosis and showed that they were absent in other lung diseases. In a series of sputum samples sent to him by a friend he was able to pick out the one non-tuberculous sputum which had been included by mistake. Although a number of people claimed that Ehrlich's technique was too difficult and proposed alternative procedures his principle was soon confirmed. It is curious, however, that certain of the minor modifications introduced have, to a large extent, robbed Ehrlich of the credit for making the detection of tubercle bacilli a practical proposition. Rindfleisch suggested he hasten the staining process by warming. Ziehl suggested the use of carbolic acid instead of analine water, and Neilsen the use of sulphuric instead of nitric acid, and all three of these trivial modifications have been incorporated in the staining technique known today everywhere as the Ziehl-Neilsen method.

In Berlin the examination of sputum for tubercle bacilli as an aid to clinical diagnosis was taken up enthusiastically and gradually extended all over the world, but even a year after Koch's discovery *The Lancet* considered that 'the question of the diagnostic value of tubercle bacilli is still in the region of probable rather than certain knowledge. Much more work will have to be done before their presence in an isolated excretion can be accepted as absolute proof of a tuberculous process.'[47]

Large increases in population, urbanization with squalid housing and wider facilities for travel made the nineteenth century the hey-day of epidemic infectious disease. Of all epidemics which swept the world during that century none was more spectacular than cholera. Until 1817 cholera had probably been confined to India but, during the following five years, a pandemic ravaged widely over Asia and as far west as East Africa. A second pandemic beginning in 1826 and lasting for ten years spread over the whole world causing frightful mortality in the

capitals of even the most civilized countries, as did yet a third and a fourth pandemic in 1846–63 and 1865–75 respectively.[48] It was during the third pandemic that John Snow, a London anaesthetist, showed by his classic epidemiological studies that whatever it was that caused cholera was swallowed, multiplied in the intestine, appeared in the cholera patients faeces and thence reached the alimentary canal of healthy persons; most commonly via contaminated water supplies. Snow's evidence seems to us today unassailable but it was not so to his contemporaries. Those who are interested can read a full discussion of the various theories as to the aetiology of cholera in a report of a special committee of the Royal College of physicians, published in 1854, but Snow's views were rejected.

After the fourth pandemic the more civilized parts of the globe remained free from cholera for a number of years during which time bacteriology and the germ theory of disease made spectacular advances. It was, therefore, as Koch put it, 'not unfortunate' that in the summer of 1883 cholera broke out in the relatively geographically convenient Egypt. The disease, which was at first confused with enteric fever, first broke out in June, rapidly spread over the most important parts of the country, including Cairo and Alexandria, and by early August was causing a weekly mortality of some 5,000 persons. The opportunity to investigate the cause of the disease was seized with most commendable alacrity by the two foremost schools of bacteriological science in the world; those of Pasteur and Koch.

Pasteur, already over 60 years of age, did not go to Egypt himself but he dispatched a formidable team of investigators, Straus, Nocard, Roux and 27-year-old Thuillier, with a specially drawn-up, nine-point memorandum on hygienic measures to preserve their own health based on the assumption that cholera was indeed a microbial disease. The work of the French Commission need not detain us, its results being relatively insignificant and it is today chiefly remembered for the tragic death of Thuillier from cholera on 18 September.[49]

Koch and his assistants arrived in Egypt in August and immediately set about a study of the morbid anatomy of the disease which convinced them that the essential lesions of cholera were confined to the intestine. Microscopy showed a characteristic comma-shaped bacillus in the mucosa of the

intestine. Turning to the contents of the bowel Koch found that in acute, uncomplicated cases of cholera this same microbe was present in enormous numbers.

But then the cholera, disappearing as suddenly as it had come, had by early September all but subsided. Koch's preliminary observations had been sufficiently promising to make further study worthwhile and he therefore made arrangements to go to India where the disease was endemic and cases always available. It was, as *The Lancet* pointed out, humiliating that 'it seems probable that the discovery of the true nature of the virus of cholera will be effected in England's greatest dependency, but not by an Englishman' and a disgrace that England, with her vast resources, should not have an institute, such as Koch's, already established in India. In India Koch quickly confirmed his Egyptian observations and went on to cultivate his 'comma bacillus' on a variety of artificial culture media and noted that 'In nutrient jelly the colonies of the comma bacilli take quite characteristic and definite form, which, as far as I have investigated, and as my experience goes is like that formed by no other kind of bacillus.' He noted that the bacillus was a strict aerobe, that it did not produce spores, that its growth was prevented by the least degree of acidity in the culture media and he introduced a new test, the 'gelatine stab' culture in which the comma bacillus produced a characteristic appearance.

But what was the evidence that the 'comma bacillus' was the cause of cholera? It seemed to be constantly associated with the disease. Koch found it by microscopy and by culture in forty-two post-mortem cases and in the stools of thirty-two cases during life. Moreover he was able to find the same organism in material from cases in distant parts of the world such as Alexandria and Toulon. He carefully examined the bodies of thirty persons dead of other types of intestinal disease, such as dysentery and typhoid fever, without finding the 'comma bacillus', nor could he find it in various samples of water or in animals. He was confident that 'the comma bacilli are the constant companions of the choleraic process, and that they are present nowhere else'. If this association was accepted there seemed, to Koch, to be three possibilities: Firstly, that cholera in some way altered the bowel and made it peculiarly suitable for the multiplication of the 'comma bacillus', but if this were so one would

expect to find some such organisms in healthy persons. Secondly, it might be that, in cholera, other bacteria assumed the comma-like shape. As Koch remarked, 'Some years ago, when bacteric investigating was yet in its infancy, one might have suggested such an hypothesis with some degree of justification. But the more the knowledge of the bacteria has advanced, the more it has become apparent, that as regards their form the bacteria are extraordinarily constant.' The third remaining possibility was that the bacillus was the cause of cholera and this Koch regarded as proved. He recognized the desirability of being able to transmit the disease to experimental animals using a pure culture, as he had done with the tubercle bacillus, but there were no animals naturally susceptible to cholera and he had not been able to produce the disease in experimental animals . . . 'Hence we must give up this method of proof.' He pointed out that the leprosy bacillus, by that time accepted as the cause of leprosy, had not been cultivated, that there were other human diseases, such as typhoid fever, which could not be transmitted to animals as well as many animal diseases to which man was immune. He saw that his 'postulates' worked out for tuberculosis could not be made to fit a number of diseases accepted as being caused by a particular bacterium and so 'a deduction can be fairly made from analogy'. Koch read a paper describing his cholera investigations to a conference held in Berlin, in July 1884, and concluded with a thorough discussion of the epidemiology of the disease which, in fact, added little to the publications of Snow over thirty years before. But he pointed out that bacteriological investigation enabled an early diagnosis to be made, even in mild cases, and could therefore be important in the control of the disease.[50]

Once again one cannot but admire the steady, unerring progress of Koch's investigation and the speed (less than a year) with which it was brought to a successful conclusion. This seems all the more remarkable when we remember that Koch was by no means the first to examine cholera material microscopically. Highly competent observers like T. R. Lewis and D. D. Cunningham had spent more than ten years in India studying cholera from the aetiological point of view. It is true that these two workers were, initially, charged with confirming or refuting Hallier's fungal theory, but their methods both of

general approach and technique lacked the crisp, sure touch of Koch. Koch started by a careful preliminary study of the morbid anatomy and histology of the disease which immediately indicated to him the site which he must search for the causative organism. Lewis and Cunningham 'messed about' examining the blood and injecting choleraic material intravenously into animals.[51] Again, even after Koch had fully described his discovery, a British commission of by no means incompetent workers, specially sent to India to examine Koch's claims, failed completely to confirm them. Two years after Koch's publication eminent British scientists, like C. S. Roy and Charles Sherrington, reported that they had found a fungus cause of cholera. The contrast between Koch's results and those of other workers in the same field is a measure of his technical superiority but it was a technical superiority amounting to genius.[52]

There was however a genuine difficulty in that other microbes morphologically identical to the cholera vibrio were found elsewhere than in cholera patients. T. R. Lewis, in 1884 assistant professor of pathology at the Army Medical College, hastened to Marseilles during the summer vacation to try to confirm Koch's findings. He failed to do so but reported the finding of morphologically identical 'comma bacilli' in the normal human mouth.[53] Finkler and Prior in the same year reported a 'comma bacillus' from the faeces of patients with diarrhoea but not true cholera.[54] But Koch had never wholly relied on the comma-shape to distinguish the true cholera vibrio and, in particular, had drawn attention to its appearance when grown in a gelatine-stab culture. Using this test there was no difficulty in distinguishing the cholera vibrio from Finkler's vibrio; the former grew only at the surface dissolving a shallow pit in the gelatine, whereas the latter excavated a deep finger-stall-like cavity along the whole needle track. This was an important step forward in clinical bacteriology for it introduced a new principle in the identification of bacteria – the biochemical reaction. Hitherto recognition had depended on microscopic appearance, behaviour with certain analine dyes, appearance of the isolated colonies and effect on experimental animals.

In October 1884 Koch began to give the first course in practical instruction in Bacteriology. He was provided with a

well-equipped teaching laboratory and the object of the course was to acquaint a large number of physicians, as rapidly as possible, with the means of making a bacteriological diagnosis of cholera. Delegations of between four to six doctors from the principal towns of Germany spent ten-day periods of practical work under Koch's direct supervision. A few students were taken from other parts of the world.[55]

The year after Koch completed his cholera studies, at the age of forty-two, he was made professor of hygiene in the University of Berlin. He, of course, continued active in research for no man ever lived more faithfully by the motto with which he headed a student essay – '*Nunquam otiosus*' (never be idle). However the next five years of his life are mainly important for his teaching and the supervision of the work of his assistants who were themselves making important discoveries.

Koch's personal research interest had centred mainly on tuberculosis since he first discovered the causative bacillus and with this he continued, despite the constant interruptions which his fame now put in his way. Almost certainly it was Pasteur's success in attenuating the microbes of anthrax and chicken cholera that inspired Koch with the desire to achieve similar success with tuberculosis. He seems to have started systematically testing a wide variety of chemical substances for their ability to inhibit the growth of the tubercle bacillus, *in vitro*, and then tested their effects on tuberculous guinea-pigs. He found numerous compounds that were effective in vitro but none had any effect on the disease in vivo.

In the course of this work Koch observed a new phenomenon – the difference in response to an injection of tubercle bacilli between a healthy guinea-pig and one which was already infected with tubercle bacilli. In the former a nodule slowly developed which eventually broke down and ulcerated through the skin and persisted to the death of the animal. In the latter, after about twenty-four hours, there was a brisk local inflammatory response followed by necrosis of the area, sloughing and rapid healing of the damaged skin. This was the first observation of what we now call 'bacterial allergy'. Koch did not grasp the significance of this observation but was intensely interested, from a therapeutic point of view, and in his investigation into the cause of this effect, he discovered what later became known as 'tuberculin'.

A filtrate of broth culture of tubercle bacilli produced the severe local reaction in a tuberculous guinea-pig although without effect on a healthy guinea-pig. A sufficient dose of 'tuberculin' would kill an infected animal, but, again, was harmless to a healthy one.

It is difficult to unravel the history of this phase of Koch's work because of the secrecy with which Koch, for some reason, surrounded it. He did not publish his account of bacterial allergy until 1891[56] – that is after he had already set the world agog with his announcement that he had developed a cure for tuberculosis. In 1890 Koch was invited to give one of the addresses at the International Medical Congress to be held in Berlin. There seems little doubt that he was unwilling so to do but Ministerial pressure was put upon him, and it seems likely that, not only was he compelled to speak but, that, for political purposes, a resounding discovery was also expected of him.[57] Koch chose to speak on 'Bacteriological Research' and considered the history and recent progress of the young science and led up, almost at the end of his discourse, to his claim that he had 'at last hit upon a substance which had the power of preventing the growth of tubercle bacilli; not only in a test tube, but in the body of an animal' in which the disease process could 'be brought completely to a standstill'. He gave absolutely no details about his therapeutic substance nor any actual experimental evidence for his claims.[58] Small clinical trials were meantime being made in various Berlin clinics and, in November, three months after his original announcement, Koch published a paper giving some details. In this he excuses himself from describing the nature of his remedy, referring to it merely as a 'brownish transparent liquid', and from giving details of the animal work on which his trials in man were based. He stated that his substance was quite harmless to a healthy guinea-pig even in a dose of 2.0 ml. In a healthy man (Koch himself) 0.25 ml produced a severe local and systemic reaction. However if the dose was reduced to 0.01 ml this had no effect on a healthy man but produced a severe reaction in a case of tuberculosis. Smaller doses produced milder reactions and, where the tuberculous lesion was readily visible, as in lupus of the skin, the striking inflammatory response in and around the actual lesions was very evident. Similar but less visible reactions occurred

around tuberculous glands in the neck, around tuberculous bone lesions and doubtless also in pulmonary lesions. In lupus this reaction, which often led to sloughing of the affected area, seemed to produce some improvement with the formation of clean scar tissue. Koch believed that his reagent could, by causing a brisk inflammatory response, cause any tuberculous lesion to heal. In pulmonary tubercle he claimed that 'cough and expectoration generally increased a little after the first injection, then grew less and less, and in the most favourable cases entirely disappeared; the expectoration also lost its purulent character and became mucous . . . simultaneously the night sweats ceased, the patient's appearance improved, and they increased in weight'.[59]

The Berlin correspondent of the *British Medical Journal* reported that excitement over Koch's remedy was 'at white heat'. At least 1,500 doctors had already come to Berlin to try to learn about it and one of Koch's assistants had no less than eight consulting rooms scattered throughout the city which were crowded with patients night and day.[60] From the first Koch was criticized for keeping the nature of his 'brownish transparent liquid' secret. There were various speculations as to the reason; it was suggested that he wishes to force the authorities into providing him with better facilities. The explanation is probably simply that Koch was worried. His reagent undoubtedly had a powerful effect on the tuberculous process but the nature of this effect had not been anything like adequately investigated. Had he been left alone Koch would never have published his discovery without the thorough investigation which was so characteristic of his previous work. The tragedy was that, as a leading article in the *British Medical Journal* put it, 'The medical world has learnt to believe that any work carried out under the auspices of Professor Koch is thorough and genuine . . .' and that there was 'a feeling of confidence in the scientific value of Koch's discovery. . . .'[61] The new remedy was at first known as 'Koch's lymph' but the name was soon changed to 'Tuberculin' and the medical world was surprised when Koch eventually disclosed its nature, that it was no more than an extract of a glycerine-broth culture of tubercle bacilli.

From the first there was general agreement about the value of tuberculin in diagnosis but doubts about its therapeutic value,

particularly in pulmonary tuberculosis. It is not necessary to follow the gradual progress of disillusion in detail, which was more protracted than it might have been for two reasons; Koch's own obstinate insistence on its value and the fact that the technique of the properly controlled trial had not then been worked out. Speaking at the British Congress on Tuberculosis in 1902 Koch said he still regarded tuberculin as 'a very effective remedy for incipient phthisis'.[62] But he insisted that its use should be restricted to 'curable' cases and should only be used in patients whose temperature was normal. Cases which were incipient, curable and with normal temperatures clearly included numerous erroneous diagnoses and a high proportion of tuberculous cases which would get better anyway, so that it is not surprising that tuberculin appeared a useful remedy. But by 1902 those physicians with most experience of tuberculosis had largely abandoned tuberculin treatment. Dr. C. T. Williams, of the Brompton Hospital, said quite flatly that its therapeutic advantages were 'nil' and that Koch's suggestion that tuberculin be used only in afebrile cases was quite unhelpful since most of these patients got better anyway. Professor Osler, more politely, said that he had no experience of the use of tuberculin 'of late' (he had given it up) and that when 'the fever has disappeared we congratulate ourselves that the patient has reached a favourable stage. The choice then between the use of tuberculin and the modern open-air treatment would, I think, be decided by the cleverest physician very strongly in favour of the latter.'

Koch does not emerge with much credit from the tuberculin story which makes a sad contrast with his earlier work on anthrax, traumatic infective diseases, tuberculosis and cholera. Indeed reading Koch's papers on tuberculin it is difficult to credit that they are by the same hand as the author of those classic contributions. Yet the observations contained in this research were of fundamental importance; the discovery of bacterial allergy and the demonstration of its value as a diagnostic and case-finding tool in man and animals which would have made the reputation of any man. The pity is that Koch was no longer interested in observation but wanted to invent a cure.

Koch's other main contribution in the field of tuberculosis was no happier than his tuberculin work. It was generally

assumed that the same bacillus caused tuberculosis in man and in cattle but, in 1898, Theobald Smith showed quite clearly that the bovine and human strains were distinct, differing in virulence for rabbits and in certain cultural characteristics. Koch accepted this and went on to maintain that the bovine bacillus was non-pathogenic for man, or almost so. He based his opinion on the fact that the human bacillus was virtually non-pathogenic for cattle, as could be shown experimentally, and that primary tuberculosis of the intestine in man was a very rare disease. He did not regard the common finding of tuberculosis of the mesenteric lymph nodes in children as suggesting that bacilli had come from the alimentary tract. He expounded these views at the British Congress on Tuberculosis in 1902 and was disagreed with by distinguished veterinarians such as Nocard and Bang as well as expert human pathologists. Again Koch's great authority was a menace. As Bang pointed out, Koch's mere opinion was likely to hamper campaigns to obtain healthy milk supplies for the public.[63]

4   The Discovery of the more Important Human Pathogenic Bacteria

The streptococcus and staphylococcus

Streptococci and staphylococci were amongst the earliest bacteria to be observed in connection with human pathology because of their abundance in various sorts of septic lesions. Yet the delineation of their exact role in human disease proved more of a problem than the isolation of many of the major human pathogens. Only limited progress was made during the 'classic' days of bacteriology and, at this point, an account of discoveries in relation to these two groups of bacteria up to the end of the nineteenth century only will be given. As we have seen, Pasteur in 1879, undoubtedly saw, isolated and appreciated the significance of both pathogenic streptococci and staphylococci but he did not follow up this work and cannot be said to have 'discovered' the organisms in the full sense of the word.

Streptococci and staphylococci have always been of particular importance to surgeons and the major discoverers in this field were surgeons by profession. A convenient starting point is the work of Theodore Billroth, professor of surgery in Vienna who published an elaborate work on the bacteriology of septic infections entitled *Cocco-bacteria Septica* which William Osler described 'as of value only as illustrating the futility of brains without technique'.[64]

Billroth was well aware of Lister's work and admitted its good results, but did not subscribe to the germ theory behind it. He accepted that bacteria were commonly present in septic lesions and described three sorts; rod-shaped 'bacteria', micrococci arranged in pairs or chains, for which he introduced the term 'streptococcus', and micrococci in masses which he called 'coccoglia'. But Billroth thought, as a result of his own researches,

that 'all the above mentioned forms belong to a plant which, seeing that is composed of cocci and bacteria, and that it is generally found in putrefying fluids, I have named cocco-bacteria septica'.[65] He described a life-cycle whereby one form changed into another, the actual morphology depending on local circumstances.

The first surgical infection the aetiology of which was eluci-dated was erysipelas and was the work of a German surgeon Friedrich Fehleisen (1854–1924). Fehleisen was an assistant in the surgical clinic at Wurzburg at the time he began the studies on erysipelas which he published between 1881 and 1883. Fehleisen although not a pupil of Koch's, was familiar with the latter's techniques and had some assistance from him when he moved to Berlin, in the middle of his erysipelas studies. Fehleisen's work was yet another triumph for Koch's methods. At the time Fehleisen began his work, although bacteria had been seen, as early as 1869, by Huter in the bloody fluid squeezed out of a puncture in erysipelatous skin, there was even doubt as to whether or not the disease was contagious. Moreover workers subsequent to Huter had claimed to see a variety of bacteria in the lesions of erysipelas, and the existence of a specific causative organism was denied. Fehleisen set himself the task of deter-mining whether or not there was a special kind of bacterium constantly associated with erysipelas and, if so, was it causally connected.

He first examined sections of skin taken from thirteen cases of erysipelas, either at post-mortem or by biopsy and found masses of micrococci in the lymphatic vessels of the skin. His first attempts to cultivate these organisms from blisters were un-successful so he resorted cultures, on nutrient gelatin and nutrient agar, of small pieces of excised skin. He found that he could isolate chain-forming cocci, with constant cultural characteristics, from all cases of erysipelas and distinguish them from the various other micrococci found in association with other forms of sepsis. Pure cultures were inoculated into the tips of the ears of rabbits which developed a spreading, erythe-matous lesion with histological appearances closely resembling erysipelas in man. A traditional belief that an attack of erysipe-las was good for a variety of human diseases, particularly various tumours, made it ethically possible for Fehleisen to prove, on

man, that the organism he had isolated did indeed cause
erysipelas. In all he inoculated seven persons with pure cultures
producing typical erysipelas in six. He noted that the seventh
patient, who was unsuccessfully inoculated twice, had previously
had several attacks of the natural disease and concluded that
he was therefore immune. He also showed that some of his
experimentally inoculated patients were immune to further
experimental inoculations.[66]

The next important contribution to the aetiology of septic
infections was made by the Scottish surgeon, Alexander Ogston
(1844–1929). Ogston was born in Aberdeen, the son of the professor of medical jurisprudence in that university. After school
and university in Aberdeen and a period of study in Germany
he qualified with highest honours in 1865. He held a number of
appointments in Aberdeen whilst he gradually built up his
reputation as a surgeon. He was appointed junior surgeon to the
Royal Infirmary in 1870 and was, by 1882, senior surgeon and
Regius Professor of Surgery in the university of Aberdeen. His
career as a surgeon was long and distinguished, including the
appointment as Surgeon-in-Ordinary to Queen Victoria, but
his bacteriological work was concentrated about the years
1880–2. Ogston himself has given a short account of his discovery of the staphylococcus in some autobiographical notes[67]
but more details can be obtained from his classic paper published in the *British Medical Journal*. Ogston had qualified in
pre-Listerian days but was one of the first to grasp the importance of the antiseptic system and germs as a cause of sepsis.
Indeed his thinking on the subject was much clearer than
Lister's, with whom he engaged in respectful but firm controversy. The examination of some pus stained with analine violet
under his students microscope and the finding of masses of
cocci stimulated him to take up the subject in detail. He built
a small laboratory behind his house, obtained the necessary
vivisection licence and, with the aid of a grant from the British
Medical Association, he obtained a good Zeiss microscope,
with an oil immersion lens and Abbé condenser and set to
work. He followed the microscopical technique of Koch, but, at
the time, 1880, Koch had not worked out his culture methods
and Ogston was forced to use the various sorts of fluid
cultures available and invent his own. The difficulties and

complexities of the pre-solid medium days are well illustrated in his papers.

Ogston first examined pus, carefully removed with antiseptic precautions, from a series of eighty-two abscesses. Thirteen were typical 'cold abscesses and in these he found no organisms, but in the remainder, of which sixty-five were acute abscesses of only a few days duration, everyone contained micrococci. Ogston distinguished two sorts, those forming chains and those 'grouped like the roe of a fish into clusters'. For the most part he found either one or the other type in a particular abscess but sometimes they were mixed. He remarked that 'sufficient evidence was not obtained to decide whether these different appearances indicated different species of micrococci; but the constancy with which chains produced only chains, and groups only groups, in the various experiments that fall to be detailed subsequently, rather favoured the suspicion of their being so'. The experiments alluded to consisted of injecting pus containing micrococci into experimental animals, as well as heated pus and pus mixed with carbolic acid, only the first of which caused abscess formation and also attempts to cultivate the organisms.

His culture methods are worth describing in some detail as an example of bacteriological technique immediately prior to the introduction of solid media by Koch. In all Ogston undertook 118 culture experiments. 'Cultivations of pus of acute abscesses gave at first the most inexplicable and contradictory results. They were grown in cells, prepared by cementing a ring of glass to the upper surface of a slide, moistening the lips of the cup so formed with a weak mixture of oleate of mercury and olive oil, scorching the cup and oil in the flame of a spirit lamp, to destroy all organisms, and dropping on to it a cover-glass, also scorched in the spirit-flame. The cells were charged with the fluid in which growth was to take place, by removing the cover-glass for an instant, and dropping on it, from a Lister's flask, a minute drop of the liquid. The liquids used were: Cohn's fluid, Pasteur's fluid, urine, acitic fluid, ovarian fluid and blood obtained from the umbilical cord of new-born infants. All liquids used for cultivation were kept for at least a month before being used, and were, during that time, repeatedly examined as to their freedom from organisms.' One is hardly surprised to hear that,

for various reasons, 'it was speedily evident that no useful or uniform results were to be obtained by these cultivations'. A second method of cultivation was therefore adopted. 'On a piece of plate-glass was placed a small bottle, capable of containing half a fluid ounce. This was covered by a small glass shade, and this again by a larger one. The shades fitted accurately the surface of the plate-glass, and allowed the entrance of air, but not of solid particles.' Having sterilized the whole by hot air the bottles were carefully filled with culture fluid. Groups of four bottles were used on each occasion, one for inoculation with pus, two for deliberate contamination and one for a sterile control. Some success seems to have been achieved but the cocci cultured from pus, although clearly different from putrefactive organisms, repeatedly failed to cause infection in experimental animals. Finally, Ogston wrote, 'I hit on the idea of growing them in eggs, where, I anticipated, they would be in almost identical conditions with those under which they grew in the bodies of animals.' The eggs were washed in 5 per cent carbolic acid and carefully inoculated under a Lister's carbolic spray. When the eggs were opened after incubation they showed no signs of putrefaction but contained masses of micrococci and a small drop of the infected egg albumen caused abscess formation in experimental animals. Ogston carefully passaged the organism through a series of eggs before injecting an animal so that 'the cocci, diluted so that (assuming an egg to contain thirteen fluid-drachms) only 1/146,016,000th of a drop of the original pus could have been injected, produced an abscess – a result inexplicable save on the assumption that they were the sole cause.'[68]

But Ogston's apparently clear-cut results were by no means forthwith accepted, chiefly because opposition to them came from a most influential and unexpected source – Lister. Ogston's paper had been published in March and, in August of the same year, Lister opened the discussion on micro-organisms in relation to unhealthy wounds at the London International Congress. Although expressing himself more than ever convinced of the importance of microbes in wound infection, he, on that occasion, thought it necessary 'to utter what seems to me a needed note of warning against a tendency to exaggeration in this direction'. Lister's whole antiseptic system was based on the exclusion of

bacteria from wounds so as to avoid infection but he thought it a very different matter to conclude that they were always responsible for inflammation and suppuration. He thought that acute inflammation was often caused 'through the nervous system'. For example, the erythema around sutures was 'indubitably brought about by sympathy'. He commented specifically on Ogston's work pointing out that he had not found organisms in the pus of 'cold' abscesses and that therefore, clearly, microbes were not the sole cause of suppuration. Lister concluded 'that micrococci are, so to speak, a mere accident of these acute abscesses, and that their introduction depends upon the system being disordered'.[69] Coming from the foremost surgeon in the British Isles, a much respected colleague seventeen years his senior, this must indeed have been painful to Ogston. In addition, came almost at the same time, a paper published in *Virchow's Archiv*, by a Dr Uskoff of Cronstadt, showing that suppuration could be induced in experimental animals by the injection of a whole variety of germ-free foreign substances of which turpentine was an example.

Ogston sat down to answer his critics, principally Lister, in a very long article which was published in the *Journal of Anatomy* in 1882. His reply to Lister was courteous in the extreme saying the Mr Lister's criticisms were to be blamed on his (Ogston's) 'failing to explain the inferences that should be drawn from the facts I collected regarding micrococci'. However, it is not necessary to follow Ogston as he demolishes Lister's arguments and restates his own case at length. He added little new to his original paper, the most noteworthy item being a name for the coccus that grew in masses, like bunches of grapes, and which he regarded as a species distinct from those which grew in chains; he named it 'Staphylococcus'.[70]

Ogston's observations did not have to wait long to be fully vindicated and the staphylococcus further characterized for, in 1884, Frederick Rosenbach (1842–1923), professor extraordinary in surgery at Göttingen, published his famous monograph on 'Micro-organisms in human traumatic infective diseases'. Rosenbach, like Ogston, followed Koch's techniques but had the advantage that the method of culture on solid media was, by then, available. Rosenbach gave Ogston full credit for his work but was able to extend it in a number of important ways.

Ogston had surmised that the organisms now known as streptococci and staphylococci were distinct; Rosenbach rapidly proved this correct by the totally different character of their growth on artificial media. He accepted Ogston's name staphylococcus, for a group of organisms which he immediately saw was not homogeneous, but of at least two sorts; one producing golden yellow colonies and the other white colonies. He named the two sorts 'Staphylococcus pyogenes aureus' and 'Staphylococcus pyogenes albus' but clearly was not dealing with the organism now known as *Staphylococcus albus*, since Rosenbach's 'albus' strains were isolated from severe lesions in man and were fully pathogenic for laboratory animals. He was, almost certainly, dealing with strains of *Staphylococcus pyogenes* of differing degrees of pigmentation. Rosenbach also cultured cocci from a considerable range of septic conditions sometimes isolating staphylococci and sometimes streptococci, thus more accurately defining their pathological significance for man. He was able to confirm Pasteur's brilliant prediction that acute osteomyelitis was indeed 'a furuncle of the bone marrow'.[71] Experimental proof that cultures of *Staphylococcus pyogenes* were infective for man was provided by the Swiss surgeon K. Garré (1867–1928) who, in 1883, rubbed a culture into the skin of his forearm giving himself a severe carbuncle.

Scarlet fever, an important scourge of the urbanized, nineteenth-century populations was, by the end of the century, widely regarded as a streptococcal disease, although conclusive proof was not obtained until well into the present century. Although there are many early descriptions of this disease it was the work of the master clinicians of the last century that gradually defined the disease, showed its relation to tonsillitis and distinguished it from other throat infections. The earliest bacteriological studies implicating the streptococcus as the causative agent are to be found in Loeffler's classic paper on the aetiology of diphtheria. Loeffler described chains of cocci from the surface of the tonsils invading the lymphatics. He cultured the organism and showed that it produced suppurative lesions in experimental animals. From that time onwards the frequent, it not constant, association of streptococci with scarlet fever was admitted by most competent observers. Another important link in the chain of evidence incriminating streptococci as the cause

of scarlet fever was the reports of numerous outbreaks of the disease associated with the drinking of milk from particular sources. The earliest of these to be reported was an outbreak in Marylebone, in 1885, investigated by W. H. Power and E. Klein. In this fine epidemiological study which Bloomfield thought 'should rank as a classic with Budd's observations on typhoid fever' the source of the milk was traced to a single cow with a diseased udder. From these lesions Klein isolated abundant streptococci, showed that they could produce a generalized disease in calves and correctly concluded that the streptococci would find the milk a good medium in which to multiply, so that, when drunk, it would 'practically correspond to an artificial culture of streptococcus'.

No one in the nineteenth century, appears to have had clearer ideas on the role of the streptococcus in scarlet fever than A. Bergè who, in a paper published in 1893, concluded that scarlet fever was a manifestation of local streptococcal infection, producing the rash by means of a soluble erythrogenic toxin. He was aware of the occasional cases of scarlet fever associated with streptococcal infection in sites other than the tonsil and realized that immunity following an attack was confined to the production of the rash only and not to streptococcal infection. But Berg's views appear to have made little impression and were forgotten only to be rediscovered some thirty years later.[72]

Diphtheria

Although, as we have seen earlier, Brettoneau was convinced that diphtheria was an infectious disease it was not until 1869 that Trendelenburg reported that inoculation of rabbits and pigeons with diphtheritic material, in some cases, produced typical false membrane. His results were confirmed and extended, in 1871, by M. J. Oertel. He inoculated diphtheritic membrane directly into the trachea of rabbits, five out of twelve of which developed typical false membrane. He also showed that intramuscular inoculation produced a local lesion but which also caused the death of the animal. Oertel was able to take material from a case of human diphtheria, passage it through four animals, and finally, inject it into the trachea of a fifth animal and produce typical diphtheritic membrane. There could therefore be little doubt that diphtheria was caused by a living

micro-organism. A number of workers courageously inoculated themselves with diphtheritic material but all with negative results. We know now that many adults are immune to diphtheria, because of the presence of antibody in the blood but, at the time, these daring experiments caused regrettable confusion in contrast with the clear-cut work of Oertel.

None the less from the 1850s onward many observers studied diphtheritic membrane searching for a causative micro-organism. Many claims in favour of various fungi and bacteria were made, the most reasonable of which was that of Laycock, in 1858, who found the yeast, *Candida albicans*, in false membrane. *Candida albicans* does indeed cause local lesions in the throats of children which are not at all unlike the membrane of diphtheria. All the earlier workers used microscopical rather than cultural methods in their studies but, in 1873, Klebs attempted to grow diphtheria organisms, in glass chambers in isinglass, as well as making microscopic studies of diphtheritic membrane. Klebs, at that time professor in Prague, was convinced that a certain fungus found on the surface of the membrane penetrated into the tissues, became a micrococcus and, in this form, was disseminated widely in the body causing fatal illness. He named the organism Microsporon diptheriticum. Klebs continued to be interested in diphtheria and at a congress held at Wiesbaden in 1883 put forward quite different ideas to those he had previously advanced. Now professor in Zürich, he had there been unable to find the Microsporon diphtheriticum of his Prague days. Instead he discovered in diphtheritic membrane small bacilli, staining well with methylene blue, and having a beaded appearance, which he suggested was due to spore formation. So confident was he that this was the causative organism that he advocated microscopic examination of methylene-blue stained material at the bedside as an aid to diagnosis. There is no doubt that Klebs saw the diphtheria bacillus but it was left to F. Loeffler, one of Koch's assistants, to prove the causal relationship. Loeffler's work makes a striking contrast with the fumblings of the earlier workers and was yet another triumph for the doctrines and technique laid down by Robert Koch.[73]

Loeffler set about his investigation exactly as had Koch in his investigation of tuberculosis and cholera. First, he made a carefully histological study of twenty-two cases of diphtheria

using methylene blue, in a slightly modified formula, now used universally under the name of 'Loeffler's methylene blue'. He found the bacilli described by Klebs in every case and also showed that they did not occur in the clinically similar, but distinguishable, 'scarlatinal diphtheria'. He noted that the bacilli were confined to the membrane and were not to be found in the internal organs of fatal cases. The bacilli were, however, always present mixed with other organisms, particularly cocci, and Loeffler appreciated, following the postulates of his master, that he must obtain the organism in pure culture and with that reproduce the disease in animals.

His first attempts to cultivate the bacillus on peptone gelatine failed but he was successful in isolating many of the cocci. These he obtained pure and found, that although they were sometimes pathogenic for laboratory animals, they never produced a disease at all resembling diphtheria. Thinking that perhaps the bacillus would not grow at less than body temperature, at which temperature the peptone gelatine was useless since it melted, he cast about for a medium which would remain solid at 37° C. He therefore took glucose-broth and stiffened it with blood serum which solidified on heating. On this medium the cocci grew well but so also did the bacilli, producing quite distinctive colonies. Pure cultures of the bacilli when injected into guinea-pigs produced a fatal disease with characteristic post-mortem appearances – local serohaemorrhagic effusion, pleural effusions and marked congestion of the suprarenal glands. As in diphtheria in man, the bacilli could only be found at the site of inoculation and not in any of the internal lesions. Loeffler therefore correctly surmised that the bacillus caused its fatal effects by liberating a powerful exotoxin. As well as fatal disease in guinea-pigs, Loeffler was able to produce typical diphtheritic membrane by inoculating pure cultures directly into the tracheas of rabbits. In attempting to assess the significance of his bacillus in human diphtheria, Loeffler took cultures from the throats of twenty healthy children and from one of them isolated an organism indistinguishable from his supposed diphtheria bacillus. Moreover he was not able to isolate his bacillus from every case of diphtheria. With typical Kochian caution he therefore felt unable to claim that the bacillus he had isolated was the undoubted, sole cause of diphtheria.

He continued to study the disease and, in 1887, was able to add a number of important new facts. He had been able to isolate the diphtheria bacillus from ten out of ten early cases and he also now recognized a new, pseudo-diphtheria bacillus which might be found in human throats. This bacillus, although morphologically very similar to the diphtheria bacillus, he was able to distinguish on the ground of morphology, cultural characters and pathogenicity. He thus showed that the diphtheria bacillus was but one of a group of organisms which we now call the Corynebacteria. He also reported the important fact that two guinea-pigs which had survived the inoculation of virulent bacilli, although being severely ill, could, thenceforth, withstand large doses of fully virulent organisms with no ill-effects. He also made partially successful attempts to isolate the toxin which he had postulated was the cause of virulence in the diphtheria bacillus.

Loeffler's work was first confirmed in 1886 by the Rumanian bacteriologist, Babes and, over the next few years, by many other workers. There was some confusion caused by the work of a young Austrian, von Hoffman-Wellenhof who soon after died of glanders. He isolated diphtheria-like bacilli not only from cases of diphtheria, but from cases of measles, scarlet fever and even normal throats. However it was soon shown that 'Hoffman's bacillus' was distinct from the diphtheria bacillus and was but another member of the Corynebacteria. The coping-stone to Loeffler's work was supplied by two of Pasteur's colleagues, Roux and Yersin who, in 1888, described a method of producing highly potent diphtheria toxin and studied its chemical properties. Their material was so toxic that 0.05 mgm sufficed to kill a guinea-pig. Roux and Yersin also stressed the importance of bacteriological examinations in the diagnosis of diphtheria and founded and developed the doctrine of the carrier state and its importance in the epidemiology of the disease. As Loeffler wrote, later in life 'Now began a new era in the investigation of diphtheria'.[74]

Typhoid fever
Thanks largely to the work of William Budd typhoid fever had come, during the 1870s, to be recognized as a contagious disease and therefore, by the growing number of believers in the

germ theory, the possibility that it was caused by a micro-organism began to be seriously investigated. There were not a few workers examining typhoid lesions microscopically, inoculating animals with typhoid faeces or blood and making crude attempts at culture work and plenty of micrococci, spore-bearing bacilli and fungi were incriminated. Klebs was there, with his enthusiasm but inadequate techniques, perhaps coming rather nearer the truth than others. All this work which it would be tedious to detail serves as a backcloth illustrating the pitfalls and technical difficulties which faced the pioneer bacteriologists and against which Robert Koch and his pupils demonstrated the truth, in this field at least, of Carl Ludwig's dictum *'Die Methode ist Alles'*.[75]

The first work to represent a real advance was that of C. J. Eberth, a morbid anatomist and pupil of Virchow. He was not a pupil of Koch's and was indeed, eight years Koch's senior. At the time of his discovery of the typhoid bacillus he was 45 years old and a professor at Zürich. Eberth published two papers, in 1880 and 1881, giving the results of his histological examination of the tissues of typhoid patients removed at autopsy. Unlike most workers, Eberth appreciated the confused bacteriology to be found in the intestinal ulcers of typhoid and therefore concentrated his attention chiefly on the messenteric lymph nodes and spleen. In sections of these organs, by staining with analine dyes and clearing in acetic acid, he was able to demonstrate masses of rod-shaped bacteria. He found these organisms in a total of eighteen out of forty cases of typhoid and never found them in the tissues of twenty-four cases of disease other than typhoid. Soon after Eberth's publication his work was confirmed by J. Coats, pathologist to the Western Infirmary Glasgow, and by G. F. Crooke, resident medical officer to the Leeds Fever Hospital, each on single cases of typhoid fever. Moreover Eberth's observations were also supported by Koch who had actually seen the organisms before Eberth.

But it was left to Gaffky, one of Koch's pupils, to show, beyond reasonable doubt, that typhoid fever was indeed caused by Eberth's bacillus and to isolate the organism and study its properties. Gaffky thought that it was highly probable that Eberth's bacillus was the cause of typhoid fever, but it was unsatisfactory that it could only be demonstrated in about half the

cases of the disease. He agreed with Eberth that it was 'judicious not to pay too much stress on the detection of bacilli in the diseased portions of intestine' and himself concentrated on the examination of the mesenteric lymph nodes, spleen, liver and kidneys of post-mortem cases. His technique was thoroughly modern; cutting sections of fixed tissue, staining with methylene blue, clearing in turpentine, mounting in Canada balsam, and examining with 1/12 in. oil immersion lens. Using this method he had no difficulty in finding typhoid bacilli in the tissues of twenty-six out of twenty-eight cases of the disease. This observation made the casual role of the organism much more certain. Gaffky made his first attempt to culture the organism in October 1881. Taking the entire spleen of a typhoid case he soaked it in mercuric chloride solution to sterilize the outer surface, and then cut into it with a sterile knife. He sliced into the cut surface with a second sterile knife and, from the centre of the spleen, fished out tiny fragments of tissue with sterile platinum wires. This tissue was streaked on to nutrient gelatine plates and left covered at room temperature. Forty-eight hours later he had a pure culture of the typhoid bacillus. He noted its morphology, staining reactions, the fact that it did not, like so many putrefactive bacteria, liquify gelatine, that it was highly motile and grew in what he regarded as a typical way in a gelatine-stab culture. He also studied the way the organism grew on potato and noted that it grew as a uniform, almost invisible film which, again, he regarded as characteristic and distinct from any other species. The character of the growth of the typhoid bacillus on potato was, in fact, to remain for some years the most certain way available for distinguishing it from other morphologically similar organisms. In only one point did Gaffky err – he thought he could see spores in the typhoid bacillus.

Gaffky attempted to isolate the bacillus from the faeces of typhoid patients but found it impossible as his gelatine cultures were invariably liquified by saprophytic organisms. He also attempted to cultivate the organism from the blood by taking blood, obtained by cupping from the cleansed abdominal wall, mixing it with molten nutrient gelatine and pouring it on to glass slides, but all he obtained were a few colonies of bacteria which were obviously different from the typhoid bacillus.

Despite his failure, he felt that, under favourable circumstances, positive blood cultures could be obtained and it is worth noting that venepuncture, such a simple and widely practiced technique today, had not, at that time, been introduced. Gaffky also attempted to infect monkeys but without success. Coming from the Koch school, as he did, Gaffky wound up his paper with a cautious discussion of the relationship of his bacillus to typhoid fever. All the criteria he would have liked satisfied had not been fulfilled, but, weighing all the evidence, he thought that 'we may regard these organisms as the cause of the disease with quite as much justice as is now the case with spirochaetae for relapsing fever and leprosy bacilli for leprosy'.[76]

None the less the importance of being able to isolate and identify the typhoid bacillus in faeces and material such as water outside the body, from the diagnostic and public health point of view, led to considerable efforts to improve upon the simple plating of suspected material on gelatine. Early work on disinfectants had shown that not all bacteria were equally susceptible to their action, and Dr Vincent of Paris tried to produce a selective medium which, whilst inhibiting the growth of saprophytic organisms, would allow the typhoid bacilli to grow. He added a small quantity of carbolic acid to broth and believed that typhoid bacilli would grow in this whilst the saprophytes would be inhibited. However, since he had no satisfactory method of differentiating the typhoid bacillus from other coliform bacteria it is difficult to assess his results. In 1890 Koch crystallized thought on the current situation when he said, at the Berlin International Medical Congress, that, in his opinion, even a skilful bacteriologist would usually be unable to identify the typhoid bacillus outside the typhoid lesions because 'constant distinctive marks are always wanting' and that recent statements about its isolation could 'only be received with legitimate doubt'. Other selective methods of isolation which were tried were those of Chantmesse and Widal, who incorporated 0.2 per cent phenol in their nutrient gelatine, Uffelmann's in which the nutrient gelatine was acidified with citric acid and methyl violet added, and Parietti's which was a modification of Vincent's carbolic broth.

The whole of this early work was hampered by the lack of any means of positively identifying the typhoid bacillus. Practically

the only test available was the character of its growth on potato. Theobald Smith struck new ground when he made an extensive study of the biochemical activity of the typhoid-colon organisms and, in 1889, showed that whereas coliform organisms fermented glucose with the production of gas the typhoid bacillus produced no gas. Three years later he reported similar results with regard to the fermentation of lactose. Kitosato's indole test was introduced about this time and it was shown that typhoid bacilli did not form indole. Klein, writing in 1894, considered that there should be no difficulty in isolating typhoid bacilli from the faeces, from the second week of the disease onwards, by direct plating on to gelatine, identifying them by their colonial appearance and observing the results of the following tests: motility, gas bubble formation in gelatine-shake cultures, action on milk and indole reaction. The identification of typhoid bacilli rested on no more certain grounds than these until the discovery and application of the agglutination reaction to bacteriological diagnosis.

In the hands of most workers the isolation of typhoid bacilli remained difficult and uncertain. Although it was soon realized that 0.2 per cent phenol-gelatine allowed the growth of organisms other than the typhoid bacillus it did at least inhibit the growth of saprophytic gelatine liquifying organisms. The first really satisfactory selective medium for the isolation of the typhoid bacillus was introduced by Drigalski and Conradi, in Germany in 1902. They added lactose, litmus and crystal violet to nutrient agar. Many of the saprophytes were inhibited by the crystal violet, and the blue colonies of typhoid bacilli could be distinguished from the red, lactose fermenting, coliforms. This medium was used extensively in Germany, America and Great Britain, and made the isolation of typhoid bacilli from the faeces, for the first time, a practical proposition. It was, however, in Great Britain at least, soon replaced by the well-known MacConkey's medium. A. J. MacConkey became interested in sorting out the various intestinal organisms whilst working as assistant bacteriologist to the Royal Commission on Sewage Disposal. His aim was to produce a medium favourable to the growth of coliforms and the typhoid bacillus but inhibiting to the gelatine liquefiers and cocci. He gradually evolved his eponymous medium during the first five years of this century,

first incorporating bile salts with glucose or lactose into nutrient agar, later adding litmus as an indicator, but finally changing the indicator to neutral red which actually stained the acid producing colonies. The value of culturing the faeces in broth containing an inhibitor of commensals prior to plating out on Drigalski and Conradi's medium was demonstrated by Peabody and Pratt in 1908. They used broth containing malachite green and this remained the standard enrichment medium until the introduction of selenite broth in the 1930s.

The principles of identification by biochemical activity were all worked out in the effort to distinguish the typhoid bacillus from the coliforms. It was observed that the fermentative activity of the typhoid bacillus, as judged by its gas or acid-production in various media, was much less than that of most coliforms. That the coliforms, but not the typhoid bacillus, produced acid from lactose was discovered by noting that the former produced curdling of milk whilst the latter did not. The test was refined by using litmus-whey, a medium consisting of milk from which most of the protein had been precipitated together with litmus to detect acid formation. A similar medium was introduced by Widal which consisted of broth containing lactose and calcium carbonate. The acid formed by fermentation acted on the calcium carbonate to produce bubbles of carbon dioxide. At first it was thought that the typhoid bacillus lacked fermentative activity but it was soon shown that although it was incapable of fermenting sucrose or lactose it could produce acid from glucose, galactose, laevulose and arabinose.

One difficulty with fermentation tests was to produce a nutritious medium that contained no fermentable carbohydrates but the one deliberately added. It was soon appreciated that ordinary broth was unsatisfactory in this respect because of its variable carbohydrate content. A method which overcame this difficulty entailed the inoculation of the broth with a coliform; after the organism had grown and utilized all the carbohydrate the broth was filtered, resterilized, and the required sugar added. Fermentative activity was also studied in a solid, gelatine medium to which sugar and litmus had been added, either 'stab' or 'shake' cultures being used. But here again the unknown carbohydrate content of the broth, used to make the nutrient gelatine, necessitated putting up a control tube of the

same batch of nutrient gelatine without added sugar. H. E. Durham described his method of testing for fermentative activity in a short paper in the *British Medical Journal* in 1898. His simple technique solved all the technical difficulties of the study of fermentation. Besides the 'Durham's tube' inserted into the medium to detect gas formation and an indicator to detect acid, Durham made up his sugar solutions in peptone water, which would support bacterial growth but which contained no carbohydrate. His technique which is, of course, still used daily in every bacteriological laboratory, enabled extensive testing of fermentative activity of bacteria to be undertaken and many valuable points of difference between morphologically similar organisms to be elucidated with the result that a large number of these organisms were more precisely identified.[77]

Pneumonia
Lobar pneumonia, or peripneumonie, had been recognized as a fairly well defined clinically and morbid anatomical entity since the masterly studies of Laennec at the beginning of the nineteenth century. It was a common disease with a very variable, but always significant, mortality. Besides lobar pneumonia the lungs were subject to a variety of inflammations some of which occurred in epidemics.

Modern investigation of lobar pneumonia began in 1874 when T. Jurgensen published a long and excellent review in which he said, referring to the aetiology of the disease, 'The assumption of a specific aetiologic agent is necessary, croupous pneumonia belongs, then, to the group of infectious diseases. . . . Not all inflammation-producing agents can cause croupous pneumonia. It takes "something" with specific characteristics – just as with typhoid.' This was a particularly perceptive remark since lobar pneumonia is by no means an obviously contagious disease. Indeed clear-cut case to case transmission is very unusual. None the less small epidemics and family outbreaks have been recorded.

In the following year E. Klebs (one almost comes to expect it) published a paper strongly supporting the thesis that pneumonia was a contagious disease and reporting attempts to cultivate the causative organism in egg white and, as usual, obtaining an organism, this time some highly motile 'monads'. In

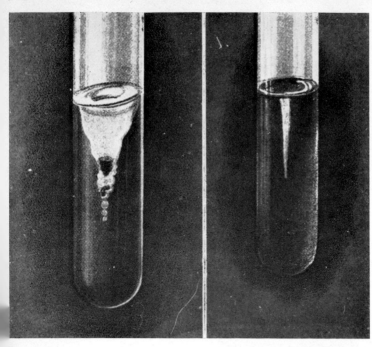

4. Appearance of growth of two different organisms in a 'gelatine stab' culture.

Woodhead and Hare 'Pathological Mycology' 1885

5. Bacterial growth on potato.

Woodhead and Hare 'Pathological Mycology' 1885

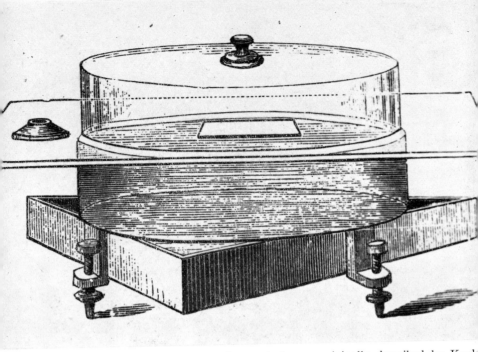

Apparatus for preparing plates of nutrient gelatine as originally described by Koch

Crookshank, 'Text-book of Bacteriology' 1896

7. Bacterial colonies growing on a nutrient gelatine plate.

Crookshank, 'Text-book of Bacteriology' 1896

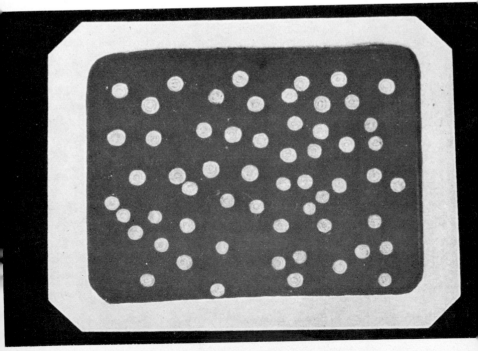

1880 Eberth described cocci from a typical case of lobar pneumonia complicated by meningitis and recognized them as different from the cocci commonly found in pyaemia. These were, almost certainly, pneumococci.

The first really significant bacteriological study of lobar pneumonia was that published by C. Friedlander in 1882. Friedlander was at that time 35 years old and morbid anatomist to the Berlin-Friedreischstain hospital. He was already suffering from pulmonary tuberculosis of which he was to die five years later. Assisting him at the time was a young Danish physician, C. H. J. Gram, who, in the course of these studies was to establish for himself a permanent place in the history of bacteriology. Friedlander's first study merely reported the presence of enormous numbers of diplococci in sections of lung from eight typical cases of lobar pneumonia. There can be no doubt that these were the pneumococcus. In the same year two physicians Leyden and Gunther reported, at a meeting in Berlin, that they had seen similar organisms in material obtained by lung puncture during life. Gunther also noticed that the cocci were surrounded by a capsule. In 1883 Friedlander published the results of a more extended study. He had found cocci in fifty cases of lobar pneumonia and reported that his colleague, Gram, had developed a staining technique which demonstrated the organisms to perfection. An unconfirmed tradition has it that Gram made the discovery of his now world-famous staining technique by accidentally spilling some Lugol's iodine solution over sections that had been stained with methyl-violet, and subsequently attempting to wash it off with alcohol. Be that as it may, Gram found that if sections were stained in methyl-violet, iodine added as a mordant and the sections then washed in alcohol the tissues lost the violet stain and the bacteria, remaining stained, showed up beautifully. He noted too that although his technique stained some organisms, such as the pneumococcus, excellently it failed to stain others such as the typhoid bacillus. Neither Gram nor the bacteriological world at large appreciated the significance of this at the time, but by the end of the century it became customary to state, in describing an organism, whether or not it could be stained by Gram's technique. Gram's career did not continue long in bacteriology. He returned to Denmark to take up a clinical career being

D

appointed professor of medicine at Copenhagen in 1900. He seems to have been a charming character and a good amateur botanist. To the end of his days it was a source of amusement to him that he should be known throughout the world by a staining technique he had devised at the age of 31. But to return to Friedlander; he managed to cultivate his organism on nutrient gelatine and, although they lost their capsules, they still stained by Gram's technique. He showed that the organisms were extremely virulent for mice but not virulent for rabbits. Here were the beginnings of confusion and controversy. Pneumonic consolidation of the lung, we now know, can be caused by more than one bacterium, although the pneumococcus Friedlander had already described is the commonest cause. Amongst these other organisms is another strikingly capsulate organism, a very short bacillus occurring in pairs and superficially very like the pneumococcus. This organism, unlike the true pneumococcus, is not virulent for rabbits and, moreover grows readily on ordinary culture medium, whilst the pneumococcus grows with difficulty. Friedlander undoubtedly amongst his typical cases of lobar pneumonia, had also an occasional case of what we now call *Klebsiella pneumoniae*. He readily isolated Klebsiella but failed to differentiate it from the commoner pneumococcus. Yet the means to do so was at hand – the pneumococcus is stained by Gram's method but Klebsiella are not. Not unnaturally other bacteriologists were interested to see cultures of the causative organism of pneumonia and Friedlander obliged by distributing cultures quite widely. Again, quite naturally, the cultures which were distributed were of the organism which grew well on ordinary media, not the delicate, true pneumococcus.

Meanwhile, in France, C. Talamon was working quite independently at the aetiology of pneumonia and a few days after Friedlander's publication, read a paper to the Societé Anatomique of Paris confirming the presence of diplococci in pneumonic lung, which he likened to grains of barley and, by introducing them directly into the lungs of rabbits, produced typical lesions of lobar pneumonia. This was a most important point since, although the virulence of the organisms had been previously demonstrated, only a septicaemic type of disease had been produced. There is no doubt that Friedlander and

Talamon had correctly identified the causative organism of lobar pneumonia but, at the same time, the pneumococcus had been discovered quite independently by two other workers in normal saliva. The findings of Pasteur, when he commenced his studies of rabies, and of the American G. M. Sternberg have already been described. Neither of these connected the organism with pneumonia or any other human disease. None the less Sternberg recognized the similarity of the pneumococcus with his salivary organism from descriptions in the literature and made a study of pneumonic exudates, in the summer of 1885, finding the pneumococcus. Visiting Koch's institute in the autumn of the same year and being shown a culture of a pneumonia organism, sent by Friedlander, he immediately recognized it as quite different.

Meanwhile another worker had appeared on the scene – C. Fraenkel. Fraenkel, as we shall see, made important contributions to the bacteriology of pneumonia but also devoted much energy to depriving Friedlander of as much credit as possible. His first contribution, a paper read at the Wiesbaden Congress, in April 1884, was unhelpful. He claimed that there were two different capsulate organisms to be found in the respiratory tract, the causative organism of pneumonia and the salivary organism which he designated the coccus of sputum septicaemia. It was not until 1886 that he conceded that the pneumococcus and the sputum septicaemia organism were identical and reported a thorough study of the cultural characteristics and animal pathogenicity of the organism. Fraenkel was both luckier and perhaps a more skilled bacteriologist than Friedlander and did not include cases of *Klebsiella pneumoniae* for his confusion. He therefore waxed somewhat caustic about Friedlander, pointing out that 'Friedlander's bacillus' was not the cause of lobar pneumonia, as indeed, the cultures which Friedlander had distributed were not. Fraenkel tried, successfully, to saddle Friedlander with his bacillus which we now call Klebsiella, and had no hesitation, in his successful textbook of bacteriology, in referring to the causative organism of lobar pneumonia as 'Fraenkel's bacillus'. He also made the vital point that Friedlander's bacillus was Gram negative whereas the true pneumococcus was Gram positive. One other worker who confirmed and in some ways extended the work of Friedlander and

Fraenkel, was A. Weichselbaum. He particularly stressed that
the pneumococcus frequently caused lesions elsewhere than the
lungs such as meningitis, endocarditis, pericarditis, arthritis, etc.,
and also made it clear that pneumonia could be caused by
more than one organism, including 'Friedlander's bacillus'
(Klebsiella) which he found in the lung exudate in 9 out of
129 cases he studied (78, 79, 80, 81).

The Neisseria

Organisms belonging to the genus Neisseria are responsible for
two important human diseases, gonorrhoea and cerebrospinal
meningitis. But, although the diseases in question have nothing
in common, the history of our knowledge of their bacteriology
can conveniently be dealt with together.

It is not surprising that gonorrhoea was one of the earliest
human diseases to be shown to be caused by a bacterium. The
disease is highly infectious and the purulent urethral discharge
contains the causative organisms in large numbers. None the less
its demonstration was not as easy as one might suppose. Able
workers had looked without success for a bacterial cause.
W. Watson Cheyne, probably the ablest of the early British
bacteriologists, and already a student of the bacteriology of
wound infections, in the late 1870s, added gonorrhoeal pus to
fluid culture media, such as meat or cucumber infusions, and
not surprisingly, isolated a variety of irrelevant bacteria.[82]

The gonococcus was discovered, in 1879, by Albert Neisser,
then a 24-year-old assistant in Oscar Simon's dermatological
clinic in the University of Breslau. The discovery, which had
eluded others, was a triumph for the first of the twin pillars of
the Koch's method – the examination of properly analine-dye
stained material under a good microscope. Neisser stained films
of gonorrhoeal pus from the male urethra with methyl violet and
had no difficulty in finding characteristic organisms in thirty-
five cases. He also saw them in the discharge from the eyes in
ophthalmia neonatorum, but failed to find similar organisms
in pus from other situations. He noted the characteristic shape
of the organism, ascribing this to its mode of cell division and
the predominantly intracellular situation of the cocci.[83] Although Neisser reserved final judgements as to the aetiological
role of the organism, *The Lancet* commented that 'the universal

coincidence with the disease certainly demonstrates their intimate connection with it'.[84]

Neisser's observations were soon confirmed all over the world but a number of confusing points were raised which took some time to sort out. Neisser correctly ascribed a characteristic shape to his organism but with this G. M. Sternberg could not agree. He claimed of course correctly, to have seen exactly such organisms in human saliva and to have cultivated them in malt extract. He further claimed that there was nothing special about the intracellular situation of Neisser's organism pointing out that Ogston had found the cocci in abscesses to be often so situated. He reproduced crude drawings purporting to illustrate their close similarity. Sternberg also claimed to have made broth cultures of the organisms from gonorrhoeal pus and an acute abscess and was unable to detect any difference between them. Unhappily experimental animals appeared immune to gonorrhoeal infection.

The work of Koch about this time had emphasized that proof of a causal role required that the disease be transmitted by pure cultures of the organism which were many generations from the original material. During the first few years after Neisser's discovery several attempts to do this were made with variable degrees of success. But it could not be said that unequivocally pure cultures of Neisser's organism had produced gonorrhoea in the human volunteers tested. The gonococcus is a delicate organism not easily cultivated and undoubtedly most of the organisms cultured from gonorrhoeal pus by the earlier workers were not that organism. It does seem possible that Leistikow, in 1880, may have successfully cultured the gonococcus on blood serum gelatine but failed to transmit the disease to a variety of experimental animals. It is satisfactory to note that practical use of the probable bacterial aetiology of ophthalmia neonatorum was made almost immediately. A Berlin obstetrician, C. Crede, began to instil silver nitrate solution into the eyes of all new born infants under his care and immediately virtually abolished what had been a very common disease.

Neisser returned to the subject with a paper published in 1882. He confirmed the constant association of his organism with gonorrhoea and claimed to have obtained pure cultures

of it on meat extract – peptone gelatine. The distinguished bacteriologists, including Koch and Ehrlich, to whom he showed his cultures, agreed that it certainly appeared as though he had obtained pure cultures of the gonococcus. Unfortunately inoculation of his cultures into man failed to cause gonorrhoea. It is very unlikely that Neisser had isolated the gonococcus on a meat extract – peptone medium but quite likely that he had grown one of the non-pathogenic Neisseria (just as Sternberg had found in the saliva) which are morphologically indistinguishable from the true gonococcus.

The whole subject was neatly tidied up by Ernst von Bumm, professor of gynaecology in Basle in 1885. He published a small book on the subject in which he reviewed the whole subject of micro-organisms in gonorrhoea. He pointed out that there were a whole group of morphologically similar organisms, which varied in colonial colour and other characteristics, which were not the cause of gonorrhoea. Von Bumm listed the whole range of situations in which the gonococcus might be found, urethra, conjunctiva, uterine cervix, Bartholin's glands, etc., and gave a good account of the pathological histology of gonorrhoea, based on a carefully built-up personal collection of material. He also successfully transmitted gonorrhoea to a woman by means of a pure culture of the gonococcus. None the less the delicacy of the gonococcus and the difficulty of culturing it artificially mitigated against the regular repetition of infection experiments, and it was not until Wertheim, in 1891, succeeded in isolating the gonococcus easily, on serum-peptone agar and infecting five out of five men that the causal role of Neisser's gonococcus was finally settled.

By 1886 it was known that meningitis might be caused by an organism indistinguishable from the pneumococcus. But, in the following year, Weichselbaum who had been one of the first to show that the pneumococcus caused meningitis, reported finding, in six post-mortem cases, 'an entirely different kind of bacteria'. In stained films of meningitic exudate and ventricular fluid he found a moderate number of cocci which, he said, 'remind one of gonococci'. He was able to cultivate the organisms on nutrient agar, show that rather large doses would kill experimental animals and that, although the cocci stained well with methyline blue, they would not stain with Gram's method.

He named the organism 'Diplococcus intracellular meningitidis'. During the next few years nothing was done to confirm the work of Weichselbaum and his organism was not even described in standard textbooks such as Fraenkel's or Crookshank's. Streptococci and staphylococci were known to cause occasional cases of meningitis but the pneumococcus was regarded as the commonest cause. The pneumococcus and the meningococcus were doubtless sometimes confused, the crucial significance of Gram's stain not, at the time, being appreciated. Moreover the pneumococcus probably was the commonest cause of sporadic meningitis and there appear to have been no considerable epidemics of meningitis in Germany during the years about 1890. In the five years prior to 1897 there was not a single case of meningitis reported as due to Weichselbaum's coccus in the Boston City Hospital. Indeed Weichselbaum's findings were not confirmed until 1895, when Jager reported the presence of similar diplococci in twelve cases of epidemic meningitis in the garrison of Stuttgart. In the following year, a Berlin paedictrician, O. Heubner, isolated Weichselbaum's organism from the cerebro-spinal fluid taken by lumbar puncture during life, grew it in pure culture and produced meningitis in a goat by intrathecal injection.[85]

Thus by 1896, although the basic facts relating to the bacteriology of meningitis were known, there was considerable obscurity concerning the relation between the epidemic and sporadic disease and the variety of organisms implicated. Therefore when, in 1896–7, there occurred a considerable outbreak of meningitis in Boston, Professor W. J. Councilman and his colleagues F. B. Mallory and J. H. Wright took the opportunity to make a thorough investigation. Their monograph, published in 1898, based on the personal study of 111 cases gives a full and excellent account of the epidemiology, clinical features, morbid anatomy and bacteriology of the disease. Little was added until the studies undertaken during the First World War. Councilman and his colleagues fully established the 'Diplococcus intracellularis meningitidis' of Weichselbaum as the cause of epidemic meningitis. That organism, and no other, was obtained from 31 out of 35 cases at post-mortem and from 38 out of 55 samples of cerebro-spinal fluid taken during life. They gave a detailed description of the cultural characters of the organism,

experimented with its viability outside the body and successfully repeated Heubner's experiment of causing meningitis in a goat.[86]

Undulant Fever

One of the few bacterial diseases whose causative organism was not isolated by Koch or one of his pupils was that of Undulant fever. This disease has many names in different parts of the world and in a history of its bacteriology, is perhaps best referred to as 'Malta fever' for it was in Malta, and under this name that the disease was known to the discoverer of the causative organism, Surgeon David Bruce of the British Army.

David Bruce was born in 1855, in Australia whither his father, a mining engineer, had gone to install equipment. He returned to Scotland at the age of 5 years and was brought up there. Leaving school early he spent a few years in business in Manchester but then decided to study medicine and entered Edinburgh University at the age of 21 years. He did well as a student and qualified in 1881. Moving south he took an assistantship in general practice in Reigate where he met a doctor's daughter to whom he became engaged. Largely because lack of capital precluded him from setting himself up in practice and being able to marry he joined the army in 1883. He got married and was posted to Malta in 1884. From this point, in discussing the work of David Bruce, it is probably more accurate to refer to 'the Bruces' for there can have been few women who participated in and helped more to further the careers of their husbands than Mrs. Bruce. She travelled with her husband wherever his duties took him, even to camps in what then might reasonably be referred to as 'darkest Africa'. As well as looking after him, which she always did most efficiently, she taught herself to be an expert laboratory technician and, as an artist, helped to illustrate her husband's papers. Some at least who knew the Bruces considered that she supplied a considerable amount of the drive behind David Bruce's very fruitful research career and that without her he would have achieved little.

Soon after his arrival in Malta Bruce became acquainted with 'Malta fever', to him a new disease which, although common

and well known in the Mediterranean area, was not recognized in England. It was a common disease in the army. Although Bruce's hospital served only 2,200 troops, in the year 1886, there were ninety-one cases of Malta fever. Fortunately the disease usually had a low mortality (none of the ninety-one cases had died) but the average stay in hospital was as long as eighty-five days. Clinically the disease was not unlike typhoid fever, its most constant signs being fever and splenomegally. In 1887 Malta fever seemed to take a more virulent form, for by July, in contrast to the previous year, nine soldiers had died of the disease. At autopsy there were no striking abnormalities except the enlarged spleen. The ulceration of the Peyer's patches, so characteristic of typhoid fever, was not found.

Bruce was able to examine five of these fatal cases bacteriologically. His success was attributable to his following closely the methods laid down by Koch. Bruce seems to have derived his knowledge of bacteriological technique from an admirable volume, published by the New Sydenham Society, in 1886, entitled *Microparasites in Disease*, consisting of English translations of many of the papers by Koch and his pupils. Bruce followed the method of Gaffky whose isolation of the typhoid bacillus is described in that volume.

From the first fatal case, a 20-year-old private of the South Yorkshire Regiment, he prepared paraffin sections, on a Cambridge rocking microtome, of fragments of spleen tissue. These he stained with watery methylene blue and by Gram's method and, on examination by means of 1/12 in. oil immersion lens and Abbe condenser, 'enormous numbers of single micrococci were seen scattered through the tissues'. Whilst on leave in the spring of 1886 he showed his sections to Dr Sims Woodhead, pathologist to the Edinburgh Royal Infirmary, who was of the opinion that Bruce had probably found the causative organism. Sims Woodhead was well qualified to give an opinion having in the previous year published a practical textbook of bacteriology. The Bruces returned to Malta in May, and he, with the aid of Dr Caruana-Scicluna, the government analyst, prepared some tubes of nutrient agar. As there were many cases of Malta fever in the wards, Bruce tried to culture the organism from blood obtained by pricking a thoroughly sterilized finger, but without success. However, in June, a patient with Malta fever

died and Bruce, at once, made a post-mortem examination. He removed the spleen, which he wrapped in a cloth soaked in mercuric chloride solution, to minimize chances of bacterial contamination and took it 'to a small room in my quarters, the door and window of which had been kept shut for some time, so as to have a still condition of the atmosphere'. Cutting into the spleen with sterile knives he inoculated eight tubes of nutrient aga by stabbing with a platinum needle charged with splenic tissue. The tubes were, in fact, inoculated within one hour of the patient's death. The cultures were incubated at 37° C. but it was not until some sixty-eight hours later that minute pearly-white colonies of bacteria could be seen growing along the needle track. Under the microscope the colonies were seen to consist of 'innumerable small micrococci . . . they are very active and dance about. . . .' By the end of July Bruce had been able to isolate apparently identical organisms from three more fatal cases. He therefore published a preliminary note concluding that, 'From a consideration of the above facts I think it will appear sufficiently proved: (*a*) That there exists in the spleen of cases of Malta fever a definite micro-organism; and (*b*) That this micro-organism can be cultivated outside the human body'.[87]

Continuing his investigations Bruce took the opportunity to conduct autopsies on three undoubted cases of typhoid fever and successfully isolated the organism described by Gaffky which was obviously different from the one he had himself isolated from cases of Malta fever. He also gave the final touch to the proof that his micrococcus was the cause of Malta fever by transmitting the disease to a monkey using a pure culture of the organism. The monkey developed a fever and splenomegally and died after twenty-one days. From the spleen Bruce again isolated his micrococcus of Malta fever.[88]

Escherichia coli

So far we have only considered the discoveries of bacteria causing disease in man which were proceeding apace in the 1880s. But the human body normally harbours a large and varied flora the importance of which, in the maintenance of health, has only been fully realized in recent years. We must not, therefore, pass over a model study of part of the normal

bacterial flora of the body made by a young German paediatrician, Theodor Escherich, and published as a monograph of 177 pages in 1886. Escherich, who was born in 1857, had followed the customary German university career as a paediatrician becoming a professor first at Graz and finally in 1902 at Vienna. During his life he made contributions in the field of diphtheria, tuberculosis, serum-therapy and tetany but is today chiefly remembered for his monograph on 'The intestinal bacteria of sucklings and their connection with the physiology of digestion'. Considering the period at which Eschrich was working this is a remarkably thorough study of a very complex piece of bacteriology. After reviewing the literature on intestinal organisms from Lewenhoek forward, Eschrich described his microscopic and cultural studies of the faeces of infants, noting the bacteria found first in the meconium and then the changes brought about by a milk diet. He also conducted autopsies on babies of different ages examining bacteriologically various parts of the alimentary canal. He made pure culture on nutrient gelatine and nutrient agar and gave detailed descriptions of the cultural characteristics of the bacteria he isolated on a variety of media, tested their pathogenicity for laboratory animals and even analysed the gas produced when they fermented a sugar. With but the crudest apparatus he attempted to make anaerobic as well as aerobic cultures and discussed his findings from the point of view of the physiology of digestion and clinical significance. Among the organisms of which he gave detailed descriptions were *Pseudomonas aeruginosa* – one of the earliest and best accounts, and of an organism universally found in the intestinal canal which he called '*Bacterium coli commune*' and which we now call, in his honour, *Escherichia coli*.[89]

5 The Scientific Basis of Immunology

Pari passu with the development of knowledge of the important part bacteria played in the causation of human disease an interest began to be taken in the processes by which the body resisted invasion by micro-organisms and by which immunity following infection was achieved. These studies received a powerful stimulus from Pasteur's demonstration of methods of producing artificial immunity to such diseases as anthrax, chicken-cholera and rabies. Pasteur himself was well aware of the many intriguing problems in the field of immunity but was indifferent to theories unless they suggested experiments by which he could test them. He thought vaguely in terms of a struggle between the host cells and invading organisms and suggested that perhaps, as happened when an organism was cultivated *in vitro*, it used up an essential growth factor or, alternatively, in the course of its metabolism, produced some chemical which inhibited further growth of the organism. Neither of these ideas can be looked upon as a very serious contribution to the subject. The first positive contributions to knowledge of the mechanisms of immunity were those of the Russian zoologist E. Metchnikoff, about the beginning of 1883. He was working on intracellular digestion of food and the origin of the intestine in invertebrates. He was living privately at Messina whose straits provided him with a wealth of suitable experimental animals. Years later Metchinkoff wrote: 'Thus it was in Messina that the great event of my scientific life took place. A zoologist until then, I suddenly became a pathologist. . . . One day I remained alone with my microscope, observing the life in the mobile cells of a transparent starfish larva, when a new thought suddenly flashed across my brain. It struck me that similar cells might serve in the defence of the organism against intruders . . . if my supposition was true, a

splinter introduced into the body of a star-fish larva, devoid of blood vessels or of a nervous system, should soon be surrounded by mobile cells as is to be observed in a man who runs a splinter into his finger. This was no sooner said than done.' Metchnikoff at once introduced a rose thorn under the skin of a star-fish larva and, after a sleepless night of excitement, found his prediction fulfilled. 'That experiment formed the basis of the phagocyte theory, to the development of which I devoted the next twenty-five years of my life.' [90]

There is no doubt that the credit for developing knowledge of this important mechanism of immunity belongs entirely to Metchnikoff but, before continuing to trace the history of his researches, it will be convenient to look at some observations made by others before him which might have formed the starting point of research into cellular immunity. It was early noted by such workers as Neisser, Koch and Ogston that pathogenic bacteria were often found within the blood leucocytes but they entirely failed to appreciate the significance of the observations. They tended to think that the bacteria had 'invaded' the leucocyte. The process of inflammation was extensively studied about the middle of the nineteenth century and the diapedesis of leucocytes and the fact that they engulphed fragments of damaged tissue or injected carmine particles was well known. In fact, one or two people had suggested that the leucocytes played a part in the body's defence against bacteria; the Dane, P. L. Panum, as long ago as 1874, and Professor C. J. Ewart of Aberdeen who, in a lecture, in 1880, said that when a healthy animal was inoculated with the anthrax bacillus 'the white blood corpuscles attacked them, and the kidneys tried to throw them off'.[91] But these suggestions seem to have fallen on stoney ground and were not followed up by their authors. Moreover the distinguished physiologist, A. E. Shafer, described experiments showing that, although a leucocyte might engulf a particle (he used the leucocyte of a newt and yeast cells, starch grains and fat globules), it was unable to digest it. Indeed it seemed to him improbable that, a cell, whose natural habitat contained all its necessary nutriments in soluble form, should engulf particulate matter in nature.[92]

Elie Metchnikoff is one of the very greatest figures in the history of medical microbiology. Born in 1845, the son of a

Ukranian landowner, he grew up an intelligent boy and a good student who studied biology at Kharkoff University, as well as wandering around a number of zoological departments in Europe. He was awarded his degree in 1867 and received a prize for his embryological researches. He was immediately appointed professor of zoology at the University of Odessa. The next eight years were devoted to rather frenzied activity; research, travel, teaching, marriage to a consumptive who died after three years, illnesses both mental and physical and two attempts at suicide. In 1875 he married again, a schoolgirl hardly half his age, and settled down to his teaching and research in the University of Odessa. In 1881 he inherited enough money from his wife's parents to live independently and, since the political situation in Russia was not to his taste, he went to live at Messina where, as has already been told, his brilliant concept of phagocytosis came to him. For about the next twenty years Metchnikoff devoted himself to immunological researches and it is the details of this period we must now follow. The latter part of his life was spent in relatively unimportant studies of the intestinal flora and problems of longevity. He died in 1916.

Immediately after his experiment with the rose thorns Metchnikoff tried inoculating various microbes into larvae and marine invertebrates and was delighted to see them engulfed by the mesodermal cells. He grasped at once the significance of the diapedesis of the leucocytes in inflammation, as no trained pathologist had done, and expounded his views to the great Rudolph Virchow who happened to be passing through Messina. Virchow saw Metchnikoff's experiments and was generally encouraging, although he pointed out that Metchnikoff's ideas ran quite counter to the prevailing teaching on inflammation. In the summer of 1883 he retired to a villa on Lake Garda and wrote his first paper on phagocytosis entitled 'Researches on intracellular digestion in invertetrate animals'. On his way to Russia he visited Professor Claus, the zoologist, at Vienna and in discussing his work with him, between them, they coined the word 'phagocytes' for these 'devouring cells'. Claus published his paper for him the same year. During the summer, in Russia, he studied the part phagocytes played in the metamorphosis of echinoderms and tadpoles but made no further studies respecting infectious disease. None the less in the autumn of the same

year he read a paper to a meeting of physicians in Odessa entitled 'The curative forces of the organism' in which he compared the phagocytes to an army hurling itself upon the enemy microbes. Yet, at this time, all was but brilliant intuition for Metchnikoff had never seen a phagocyte hurl itself upon an invading microbe. But this deficiency he was soon able to make good.

Metchnikoff recalled noticing that the transparent water flea, Daphnia, was sometimes diseased and, being transparent, admirably suited to microscopic observation of its blood cells *in vivo*. In the autumn of 1883 he found some Daphnia in an aquarium which were infected with a yeast-like fungus which he named Monospora bicuspidata. The yeast spores were swallowed by the Daphnia, made their way through the wall in the intestine, proliferated in the body cavity and, in about two weeks, killed the Daphnia. Metchnikoff's observations on this disease of Daphnia are worth considering in some detail and recounting as far as possible in his own words for this paper really laid the foundation of the cellular theory of immunity. He wrote: 'Hardly has a piece of the spore penetrated into the body cavity, than one or more blood corpuscles attach to it, in order to begin the battle against the intruder . . . the blood cells fasten so tight to the spore that they are only seldom broken free by the blood-stream . . . when many spores are in the body cavity at the same time, such a large number of blood cells surround them, that the whole area appears highly inflamed, so far as one can speak of inflammation in the vesselless animal . . . (The blood cells) may unite occasionally into a more or less extensive plasmodium (a so-called giant cell).' After a while the ingested spores could be seen to be undergoing changes which he regarded as digestion. 'From what has been said above, it is evident that spores which reach the body cavity are attacked by blood cells, and – probably through some sort of secretion – are killed and destroyed. In other words, the blood corpuscles have the role of protecting the organism from infectious materials.' He also noted that although the blood cells could destroy the fungus the converse was also true and several times he observed 'a blood cell full of parasites rupture before my eyes, setting the fungus cells free again'. He also noted that the blood cells were not the only phagocytic cells in Daphnia but

that fixed isolated convective tissue cells played a similar role whereas other tissue cells such as the heart muscle did not. He commented the phagocytes 'seem, therefore, as the bearers of nature's healing power, which has been known to exist for a long time, and which Virchow first placed in the tissue elements. The whole course of the Daphnia disease fits in with the basic thoughts of this master of cellular pathology. . . .' He also made a critical comment about earlier workers, naming R. Koch as an example, who had concluded that bacteria penetrate leucocytes and there multiply. It was much more likely that they had been engulfed by the phagocytes.[93]

Following his brilliant study of infected daphnia Metchnikoff was naturally anxious to test the validity of his theory in higher animals. To start with he chose probably the best studied of all infectious diseases – anthrax. He was soon able to show that the leucocytes of the rabbit engulfed anthrax bacilli and, further than that, that the leucocytes of an artificially immunized rabbit took up anthrax bacilli avidly but those of a non-immune animal hardly at all. These observations he published in 1884 and in the next few years extended his observations to include streptococci, the spirochaetes of relapsing fever and tubercle bacilli and was able to show the importance of phagocytosis in defence against all of them.

In 1891 Metchnikoff gave a series of lectures at the Pasteur Institute under the title of the 'Comparative pathology of inflammation' which were published the following year and constitute one of the great books of medicine. In these lectures Metchnikoff stressed the value of a biological approach and a consideration of evolutionary significance in problems of pathology and, also, the fact that the essence of a complex phenomenon like inflammation might be better grasped in simple animals. He then traced the evolution of the reaction of living organisms to parasitic invasion or injury and, in so doing, put his finger on the significance of inflammation which, despite the most extensive research, had utterly eluded numerous distinguished pathologists. Inflammation was a reaction for defence and repair of damaged tissue carried out by phagocytic cells. In the two-layered larva of a sponge the endoderm was, for purposes of nutrition, phagocytic. In the adult sponge a third layer, the mesoderm, had been developed from the endoderm

and consisted of amoeboid cells with phagocytic powers. These cells no longer played a part in the nutrition of the organism but were for defence purposes. He demonstrated that the introduction of foreign bodies, bacteria or red blood cells invariably led to their being surrounded by the mesodermal phagocytes and often digested. At this level in the animal kingdom the phagocytic cells were connective tissue cells, not blood cells, for the sponge had no vascular system.

At a higher level, in earthworms, which had a primitive vascular system, the phagocytic cells were the endothelial cells of the body cavity and perivisceral tissues. The blood vessels played no part in the reaction to parasites or injury. Even in primitive vertebrates, such as the embryo of the Axolotl, whose fins were devoid of blood vessels the mesodermal connective tissue cells reacted just as in an avascular animal. If one went one step higher and tested the reaction of Triton larvae, which did have fins with blood vessels, the amount of dispedesis of leucocytes was insignificant and the inflammatory reaction still essentially one of connective tissue cells. But at a higher level still, in tadpoles with very vascular tails, diapesis became marked and constituted the major part of the inflammatory response. He wrote that 'the *primum movens* of inflammation consists in a phagocytic reaction on the part of the animal organism. All the other phenomena are merely accessory to this process, and may be regarded as means to facilitate the access of phagocytes to the injured part'.[94]

It is difficult to appreciate the greatness of Metchnikoff's concept without reading the writings of eminent contemporary pathologists. One needs to feel their hopeless flounderings and utter failure to grasp what inflammation was about in contrast with Metchnikoff's simple, elegant ideas. Many more observations were brought forward in support of his ideas. He recognized the various sorts of leucocytes in higher animals noting that there were two main phagocytic blood cells, the polymorphonuclear leucocyte and the macrophage, and that they tended to attack different sorts of microbes. He was well aware of the other fixed phagocytic cells of the body, derived from the endothelium of vessels, such as the Kupffer cells of the liver and the sinusoideal cells of the lymphatics. He adduced evidence that the inflammatory exudate, apart from the phagocytes, had

little bactericidal effect and drew attention to the interesting difference in reaction, between an immune and non-immune guinea-pig, to a subcutaneous injection of a pathogenic vibrio. In the former there was little exudation but active phagocytosis whereas in the latter the exudation was abundant but the phagocytosis ineffective.

Metchnikoff's phagocyte theory was almost immediately attacked and, to his surprise, by the pathologists. He had supposed that those who had spent so much time working at the minute details of the inflammatory process would have been well prepared to appreciate the significance of his ideas. The earliest critic was Paul von Baumgarten, professor of pathology in Konigsberg, but he was followed by E. Ziegler, professor of pathology in Freiburg, who was the author of a well-known textbook of pathology in which a particularly good account of the inflammatory process was given, including the part played by the leucocytes in engulfing tissue debris. Metchnikoff himself had used this very source for most of his knowledge of inflammation.[95] The objections of the pathologists were not well-founded nor did they suggest alternative ideas. It was just that Metchnikoff's views seemed too vitalistic and teleological. Metchnikoff had no difficulty in disposing of their objections. He pointed out that the pathologists were just not thinking like biologists and that, in fact, phagocytes and their role in defence fitted in very well with Darwinian evolution through natural selection and the struggle for existence. The frequent failure of leucocytes to eliminate invading microbes showed that the inflammatory reaction was 'not yet perfectly adapted to its object'. Metchnikoff does not seem to have been particularly perturbed by the attacks of the morbid anatomists.

The years 1885 to 1888 were, for Metchnikoff, an unsettled period and he finally decided that he could not continue to work in Russia. In 1887 he attended the International Congress on Hygiene at Vienna with the object of meeting workers in the field of infectious disease and also looking for some department where he might settle in a sympathetic atmosphere and continue working in what was now his chosen field. He investigated possibilities in Wiesbaden and Munich only to find them unsuitable but, visiting Pasteur, in Paris, where the new Pasteur Institute was being built, he was favourably received. On his

way back to Russia he called on Koch in Berlin to show him his preparations and discuss his ideas on phagocytosis. Metchnikoff left a full account of this visit and there seems no doubt that Koch treated him extremely rudely and ended by saying that he did not care whether the organisms were inside or outside the cells. Koch's pupils who, only the day before had expressed themselves convinced by Metchnikoff's preparations, now ranged themselves alongside their sceptical master. Metchnikoff decided to settle in Paris, although he would have preferred a quiet little university town to a great noisy city. Pasteur received him most hospitably and allotted him two rooms on the second floor of the new Institute. He was soon so much at home that he said that their was only one place for which he would leave the Pasteur Institute – the neighbouring cemetery of Montparnasse.

Metchnikoff had had but little regard for the attacks of the pathologists on his phagocyte theory but about this time criticisms came from another quarter – the bacteriologists. But these workers were of a different order and sought, by experiment, not only to overthrow the phagocyte theory but to offer better explanations for the phenomena of immunity and, as Metchnikoff himself admitted, this led to discoveries of the greatest value.

This new theory of immunity, the 'humoral' theory, was based on the observation of several workers that the blood of normal or artificially immunized animals was capable of destroying bacteria *in vitro* and postulated that microbes were killed by chemical substances in the blood and, at most, the phagocytes digested the dead organisms. Although not the earliest, the most careful work along these lines was that by G. H. F. Nuttall published in 1888. Nuttall, at the time, was 26 years old and working as an assistant in Professor C. Flugge's institute of hygiene in Göttingen. Nuttall mixed various species of bacteria with the defibrinated blood of different animals which he maintained at $37°$ C. for some hours. By plating out on nutrient gelatine, he was able to demonstrate a remarkable reduction in the number of viable organisms. He also showed that the bacteriocidal power of blood was lost after prolonged standing or heating to $55°$ C. Two of Koch's assistants, Behring and Nissen, extended this work showing that although blood was

bactericidal for some organisms it was without effect on others. They were also the first to draw attention to the remarkable bactericidal effect of the blood of a rat on the anthrax bacillus, to which the animal is naturally immune. The converse absence of bactericidal properties was found in the highly susceptible guinea-pig. Hans Buchner, in 1889, showed that the bactericidal agent of the blood was present in the cell-free blood serum. These admittedly striking observations seem to have made a profound impression upon contemporary bacteriologists and immediately there sprang up, one cannot but suspect as much on personal grounds as scientific, a rival school of immunology ascribing resistance to infectious disease to chemical substances in the blood serum rather than phagocytic cells.[96]

Immunity was a subject for some comment at the 10th International Congress of Medicine in Berlin in August 1890. Koch, in his famous address, in which he first hinted at his supposed cure for tuberculosis, dismissed the subject very shortly, but with his usual authority, saying, 'it is becoming more and more evident that the view which for some time held the foreground, that we had to deal with purely cellular processes, with a kind of struggle between the invading parasite on the one hand and devouring phagocytes on the other is steadily loosing ground, and that here also it is most probable that chemical processes play the chief part'.[97] Lister, however, after describing Metchnikoff's observations, relating them to some of his own and showing their general relevance to surgery said, 'various objections have been urged against Metchnikoff's views; but so far as I am able to judge he has met these effectively by his masterly researches; and his observations have been confirmed and extended by several independent investigators'.[98] The discussion at this Congress elicited some correspondence in the columns of *The Lancet* under the title of 'The Leucocyte as the Surgeon's Friend' and even so perceptive and able pathologist as A. A. Kanthack, later professor of pathology at Cambridge, quite denied the phagocyte theory. But he was answered by A. A. Ruffer, later to make a name for himself as a paleopathologist, but who was, at that time, Metchnikoff's foremost English supporter.[99] So interested was Lister in the problem that he made arrangements for a whole meeting of the section of bacteriology at the forthcoming London Congress of Hygiene

in 1891 to be devoted to a discussion on immunity. But before considering what Metchnikoff's wife called this 'vertiable tourney of opinions' we must give an account of a most remarkable discovery emanating from Koch's department which, despite Metchnikoff's firm conviction of the validity of his phagocyte theory, was to shock him by its apparent contradiction of the cellular immunity. In December 1890 C. von Behring and S. Kitasato published their discovery of antitoxins.[100]

The discovery of antitoxic immunity was one of the greatest discoveries in the history of medicine and the credit for it belongs almost entirely to Emil Behring. Behring was born in East Prussia in 1854, studied at the Army Medical School in Berlin graduating in 1878. During the next ten years his experience was varied and, in 1888, he was appointed a lecturer at the Army Medical School in Berlin. The following year he joined Koch's department. Behring had become interested in the action of iodoform, then the most frequently used antiseptic dressing in the German army. He supposed that it might act by neutralizing the toxic products of septic bacteria as much as by direct antibacterial action. At that time much work was being done on the toxic chemical products of putrifactive bacteria – the ptomaines, and Behring found that iodoform did indeed reduce their toxicity. About the time Behring joined Koch, Roux and Yersin had demonstrated that the lethal effects of the diphtheria bacillus depended upon the liberation of a soluble toxin and this disease, no doubt, seemed a suitable model on which to extend his observations. He wasted no time. It should be noted at this point that the possibility of acquiring immunity to diphtheria had already been demonstrated by F. Loeffler who was also working in Koch's laboratory. It just happened that one of Loeffler's diphtheria infected guinea-pigs had, after severe illness, survived and Loeffler had shown that the animal was now immune. At the same time, C. Fraenkel, also in Koch's laboratory, was trying to discover a method of attenuating diphtheria cultures to produce a vaccine. He, however, seems to have worked quite independently of Behring – one of the hints the delver into bacteriological history not infrequently gets that Koch's institute was not the happiest of ships.

Behring took broth cultures of the diphtheria bacillus and tried in various ways to render them harmless for guinea-pigs.

His work with iodoform had suggested the possible value of iodine compounds and he did find that guinea-pigs would survive inoculation with cultures containing 1 in 500 parts of iodine trichloride, and were subsequently immune to diphtheria. He appreciated that the immunizing agent was a soluble product in the culture rather than the bacteria themselves. He showed that the pleural exudate of an animal dead of diphtheria was toxic, but contained no diphtheria bacilli, and that repeated small doses of this exudate also rendered guinea-pigs immune. He further demonstrated that sub-lethal doses of diphtheria toxin had an immunizing effect. Reflecting on the mechanism of this immunity he first considered that it might be a sort of tolerance, analagous to the drunkard's for alcohol or the addict's for opium, but this hypothesis did not seem to fit the facts. He then thought that immunity might depend upon some property of the blood. He tested this in a peculiar way for the result of his experiment was open to several interpretations. He knew that rats are naturally resistant to diphtheria, so, taking one he injected into its peritoneal cavity a large quantity of diphtheria toxin. Three hours later the blood of this rat was injected into a guinea-pig which showed no symptoms of intoxication. If he did a similar experiment with an animal naturally susceptible to diphtheria the guinea-pig became ill. Much more important, he was able to show that the blood of immune guinea-pigs had some power to neutralize diphtheria toxin *in vitro*, although it had no harmful effect on the diphtheria bacilli themselves.

It seems likely that the rabbits and guinea-pigs which he had been able to immunize, and whose blood contained the factor which neutralized diphtheria toxin, were not very convenient for making a satisfactory demonstration. His test for 'antitoxin' was its ability to protect a diphtheria-infected guinea-pig and his immune animals were not so very immune, nor could he obtain easily the relatively large quantities of rabbit-immune serum required to protect an animal the size of a guinea-pig.

Fortunately another admirably adapted experimental model was at hand. Also working in Koch's laboratory was a Japanese bacteriologist, A. Kitasato, who had, only a year before, isolated the causative organism of tetanus and shown that this organism also produced its lethal effect by means of a soluble toxin. The

luck lay in the fact that a tiny animal, the mouse, was susceptible to tetanus. Behring had made his diphtheria discoveries by the summer of 1890 and, quickly changing his experimental model, he had, with Kitasato's help, by autumn shown that rabbits could be immunized against tetanus in the same way as against diphtheria. Now, relative to the size of a mouse, he was able to get large quantities of immune serum and was able to show that 0.2 ml of immune rabbit serum would protect a mouse from a lethal dose of tetanus toxin. The immune serum would also cure an already tetanus inoculated mouse and neutralize tetanus toxin *in vitro*.

The results of his work with tetanus, but not the method of immunizing rabbits, and the statement that his findings were also true for diphtheria Behring published, in a joint paper with Kitasato, on 4 December 1890. The following week he alone published an account of his work with diphtheria. In effect the diphtheria study illustrated the principles of antitoxic immunity whilst the tetanus work made a more convincing practical demonstration. The authors added a quotation from Goethe which seemed remarkably apt: '*Das Blut ist ien ganz besonderer Saft*'.[101] Besides his work on antitoxic immunity Behring, sometime in 1890, with the aid of his colleague Nissen, made another observation, of little practical importance, but great significance for the future development of immunological theory. They tested the bactericidal effect of the serum of animals vaccinated against various organisms. Mostly they found that the bactericidal effect had not been increased by vaccination but with one organism named by Gamaleia, a friend of Metchnikoff's, *Vibrio Metchnikovi*, the serum bactericidal effect was considerably increased. They concluded, not unreasonably, that the animals acquired immunity was to be attributed to the bactericidal substances produced in the blood by vaccination.

A few months more work sufficed for Behring to devise immunization courses which rendered guinea-pigs highly immune to diphtheria and, at the Congress of Hygiene in London in 1891, he was able to state that 'the blood of highly immunized guinea-pigs possesses outside the body the power of destroying the diphtheria poison, and further that guinea-pigs can be rendered immune by intraperitoneal injections of the blood of immunized animals, or, if infected, can be cured'.[102] The trial

of antitoxin in the treatment of human diphtheria was not long delayed; the first child to be so treated was in Berlin on Christmas night in 1891. Manufacture of diphtheria antitoxin began on a commercial scale in Germany in 1892. Production began at what is now the Lister Institute in London in 1895. Nuttall, writing more than thirty years later, commented that 'Rarely in the history of scientific discovery have the results of laboratory researches been followed so rapidly by their practical application, and few indeed are the workers in the domain of medical science who have in their lifetime seen comparable benefits accrue to mankind as a direct consequence of their labours'.[103] Behring's subsequent career was prosperous. But 36 years of age at the time of his great discovery he was, in 1893, given the title of Professor and in 1895 called to the chair of Hygiene at Marburg University. Here he developed great laboratories, which became known as the 'Behringwerk', designed to forward the practical application of serum-therapy. Behring himself had a financial interest in serum and vaccine production (he developed a vaccine against tuberculosis in cattle) and became a well-to-do landowner. Many honours came his way; in 1901 he received the patent of nobility becoming Emil von Behring and in the same year was the first recipient of the Nobel prize in medicine, four years before his master, Robert Koch. His later years were clouded by bouts of depressive illness and he died, in 1917, of pneumonia.

Despite the powerful reinforcement provided by Behring to the humoralists the discussion at the London Congress of 1891 was not disastrous to the phagocytic-theory. Those best qualified to give an opinion saw no reason why there should not be other mechanisms in immunity besides phagocytosis, which were not at all rendered obsolete by the discovery of antitoxins. Roux remarked that 'the theory of immunity propounded by Metchnikoff did not exclude the possibility of their being other means of protecting the organism, but it affirmed that phagocytosis had a wider sphere of action and was more efficacious than any other.[104] Metchnikoff, who himself was received with cheers, admitted the importance of Behring's discoveries but maintained that antitoxic immunity of the type they described for diphtheria and tetanus was the exception, not the rule. In this, experience up to modern times would agree with him for in no

other important disease of man has antitoxic immunity been found to be the mechanism of resistance. None the less Metchnikoff, at that time, could not know this and to insist that phagocytosis was more important than humoral immunity (on the evidence today we cannot say) was unscientific. Metchnikoff returned to Paris feeling that the discoveries of Behring hung like the sword of Damocles above his phagocyte theory. This attitude was unfortunate and henceforth he made no significant contribution to the science of immunology. All his ingenuity, experimental and dialectical, during the next decade was devoted to fitting every new observation in immunology into a broad picture of his phagocyte theory.

The next important advances in the science of immunology came from the study of 'Pfeiffer's phenomenon'. Richard Pfeiffer, the son of an East German clergyman, was born in 1858. His family's resources would have precluded a university education but he was fortunate in obtaining a place in an institution which trained doctors for the German army and, further lucky in that the enlightened policy of the army medical department allowed promising young men to be seconded into civil research establishments. Thus Pfeiffer, at the age of 29, was sent to work under Koch in Berlin. Unlike some pupils, he got on well with Koch and was rapidly promoted within the institute to director of the scientific section, in 1891. Pfeiffer's main contribution to science was made at Koch's institute, in the years before 1899, when he was appointed professor of hygiene and bacteriology at Königsberg. In 1909, at the age of 51, he became professor at Breslau where he remained until his retirement in 1926. Amongst many honours he was elected to the foreign membership of the Royal Society. His long retirement was passed during a difficult period of German history and he eventually died, in 1945, in Russian-occupied East Germany at the age of 87 years.

Pfeiffer's work was largely in the field of immunology centring about the phenomenon which became known by his name but he, amongst other contributions, was the first to isolate *Haemophilus influenzae*, in the course of which blood-agar was introduced to bacteriology, and he made a partial analysis of the factors required for the growth of that organism. He also isolated and characterized *Neisseria catarrhalis*.[105]

In 1889 Pfeiffer showed that it was possible to distinguish, immunologically, between two morphologically similar organisms, *Vibrio cholerae* and *Vibrio Metchnikovi*, by cross-immunization in guinea-pigs. Both organisms are pathogenic for that animal. Immunization with the homologous organism protects but the protection is quite specific conferring no immunity against the other vibrio. It was the application of this principle during the study, under Koch, of the epidemiology of a cholera outbreak in Hamburg, in 1892, which led Pfeiffer, with his colleague, V. I. Issaell, to discover what became known as 'Pfeiffer's phenomenon'. They found that if cholera vibrios were injected into the peritoneal cavity of a normal guinea-pig the organisms retained their characteristic shape and motility and multiplied freely until the animal died. But, if injected into the peritoneal cavity of a previously immunized guinea-pig, the outcome was quite different. If a small sample of peritoneal fluid was removed some minutes later, and examined microscopically, it was found that the vibrios had lost their motility, had become rounded, less refractile, clumped in small groups and showed less affinity for stains. Some time later a sample of peritoneal fluid would show numerous leucocytes actively phagocytosing the remains of the vibrios. These very remarkable findings, published in 1894, immediately attracted widespread attention and a mass of work designed to elucidate their underlying mechanisms led to advances in immunology comparable in importance to Pasteur's attenuation of anthrax bacilli or Behring's discovery of antitoxins.

Metchnikoff, in 1895, added a most significant observation. He showed that if a few drops of peritoneal exudate from a normal guinea-pig, but containing leucocytes, of course, was added to vibrios mixed with immune serum in a test-tube, the typical transformation of the vibrios seen in Pfeiffer's phenomenon took place. At that time, Metchnikoff had working in his laboratory a 25-year-old Belgian, Jules Bordet, and it was largely due to his studies that Pfeiffer's phenomenon exerted such a profound influence on the development of immunology. Bordet set out to examine the relative importance of leucocytes and serum factors in the protection of experimental animals against vibrio infections. 'Indeed, it is only necessary to add to non-immune serum which is weakly bacteriocidal, a small

quantity of anticholera serum, unheated or previously heated
to 60° C. (this latter removes its toxic properties for the vibrio),
to induce in it a very marked bacteriocidal power. Thus two
fluids, weakly bacteriocidal when separated, form together a
mixture which is strongly antiseptic.' Moreover, unhappy discovery to make in Metchnikoff's department, 'the bactericidal
power can be observed just as easily whether the serum contain
cellular elements or have been deprived of these elements . . .'[106]
The paper describing this work, published in 1895, was but the
first of many accounts of fundamental discoveries in immunology which were to come from Bordet during his long life.
In 1900 he returned to Belgium as director of an anti-rabies
institute, in 1907 he was made a professor in the Free University
of Brussels where he remained until he retired in 1935. In 1919
he was elected to foreign membership of the Royal Society and
received the Nobel Prize for his work in immunity. He remained mentally active and interested in science to the end of
his long life and Oakley[107] has told how, when Bordet was over
80, he was subjected by him, after reading a scientific paper, to
the most penetrating questions he had ever experienced – and
the subject of Oakley's paper was one on which Bordet had
never worked! Much more of Bordet's work will have to be
considered in its chronological place.

Meanwhile, contemporary with, and even before Bordet,
Pfeiffer's phenomenon was being investigated by Max Gruber
(1853–1927), at that time professor of hygiene in Vienna.
Gruber claimed later in life, and there seems no reason to doubt
him, that he discovered the specific agglutination of cholera
vibrios by antiserum in 1894. In that year he was joined by a
young Englishman, H. E. Durham, son of the senior surgeon
to Guy's hospital. Durham was 28 years old and after a distinguished student career, prolonged by two years of zoology, had
qualified and obtained a Gull studentship with which he went
to work in Gruber's department. At Gruber's suggestion
Durham joined him in the investigation of Pfeiffer's phenomenon and soon the main points concerning specific agglutination
by antisera were discovered. But in May 1895 Durham had to
return to England because of the death of his father, and, later
in the year, he demonstrated the phenomenon of agglutination
in London medical circles. Gruber, perhaps over-generously,

allowed Durham to publish a short note on the subject in January 1896 and he himself gave an account of the matter in the February of the same year. Durham's paper was entitled 'On a special action of the serum of highly immunized animals, and its use for diagnostic and other purposes'[108] and he acknowledged that the work was undertaken 'at the suggestion and under the guidance of Professor Max Gruber of Vienna to whom my best thanks are due'. Durham described the essence of the phenomenon in twenty-one brief statements. He showed that agglutination was a general phenomenon occurring with variety of organisms, that it was not absolutely specific, related organisms cross-reacting to some degree. But all of nineteen strains of the typhoid bacillus reacted with typhoid antiserum but none with a colon bacillus serum. The colon bacilli were by contrast shown to be a heterogonous group, not all strains reacting with a single serum. Meanwhile, in March 1896, Gruber was joined by A. S. F. Grunbaum, an Englishman despite his name (he found it necessary to change it to Leyton during the First World War), who was at once set to work to investigate the value of the agglutination reaction using patients serum and known typhoid bacilli in the diagnosis of that disease, an obvious corollary to Gruber and Durham's work so far. However, the Frenchman Ferdinand Widal, to whom the same idea of adapting Durham's discovery to the diagnosis of typhoid occurred independently, anticipated Grunbaum in publication.

Widal's paper appeared in June 1896, just over six months after Durham's original communication. Widal's technique consisted of adding the patient's serum to a broth culture of the typhoid bacillus, in the proportion of 1 in 10 or 1 in 15, and, after twenty-four hours' incubation, observing agglutination with the naked eye. Grunbaum, however, used a microscopic technique in which agglutination could be seen within thirty minutes. The great importance of this agglutination reaction in the diagnosis of typhoid fever was immediately appreciated. In August 1897, at the sixty-fifth annual meeting of the B.M.A. in Montreal, Widal was able to report that there were records of the test being done in several thousand cases. R. C. Cabot of Boston had done the test in 1,826 suspected cases of typhoid fever and had found it positive in 1,744 or over 95 per cent.

A. E. Wright, at the time 36 years old and Professor of

Pathology at the Army Medical College at Netley, made three important contributions to the clinical use of the agglutination reaction. He showed that a suspension of heat-killed typhoid bacilli could be used instead of the living broth culture; he adapted the reaction to the diagnosis of Malta fever, and introduced a technique for doing the test in capillary tubes. A good method, which required only the smallest quantity of serum, was very necessary since venepuncture was not at that time an established technique. The microscopic test was usually performed using a few drops of capillary blood on a slide. The blood could be allowed to dry and reconstituted by adding a few drops of water. Another popular alternative which was regarded as more satisfactory was to raise a skin blister with cantharides and collect the blister fluid for the test.

It was soon appreciated that care with a number of technical points was essential if the test was to be reliable; it was found that not all strains of typhoid bacilli acted equally well and that, unless the culture was grown at the correct pH, spontaneous agglutination was likely to occur. False positive reactions with serum from non-typhoid cases were common if the serum dilution was insufficient and it soon became the practice to set up dilutions ranging from 1/10 to 1/100. To be regarded as definitely positive the serum had to cause complete agglutination of the organisms in thirty minutes when diluted 1/30. It was also noted that the serum of typhoid patients, and indeed some normal persons, could agglutinate a variety of other bacteria, sometimes to quite high dilutions. But properly conducted the agglutination reaction proved of immense value in the diagnosis of typhoid fever and rapidly became an important part of the work of the clinical pathologist. In the clinical laboratory of St Thomas's Hospital, during the year 1898, 1,664 specimens of various kinds were examined, of which 175 were agglutination reactions, this being, except for the histological examination of operation specimens and the examination of throat swabs for diphtheria bacilli, the most frequent investigation undertaken.

Until the First World War some form of microscopic technique of the agglutination reaction was usually employed, even though there were a number of disadvantages. For instance,

since the observation of loss of motility formed part of the test, it was necessary to use a living broth culture which was both dangerous and difficult to standardize. The variability inherent in the method made it unsuitable for the quantitative measurement of the amount of agglutinin in the serum, that from any one patient giving irregularly different figures from day to day. In 1906 G. Dreyer, then working at the Staatens Serum Institut, Copenhagen, introduced his technique of making serial dilutions of serum in small test-tubes to which were added standard suspensions of selected, sensitive, killed organisms and the results read macroscopically. This technique was published in English in 1909, but it was not until the First World War that it superseded the microscopic technique. The importance at that time of being able to make a diagnosis in patients with typhoid fever who had received injections of T.A.B. vaccine made essential an accurate quantitative technique which could reliably determine a rise in titre of agglutinin as the disease progressed. Dreyer had been appointed Professor of Pathology at Oxford in 1907 and, during the First World War, he and his colleagues devoted much attention to the technique of serological diagnosis in enteric fever, defining its use and limitations with a thoroughness not achieved before, and issued standardized suspensions of organisms for use in the armed forces.

The confusion in the serodiagnosis of enteric fever by the techniques in use during the First World War was so great that serologists of vast experience were of the opinion that 'the Gruber-Widal reaction has lost its practical value, in consequence of the antityphoical inoculation'. However, the investigation of this difficulty not only resulted in a clarification of the situation with regard to the diagnostic Widal test and its restoration to a place of importance in the diagnosis of enteric fever, but also to one of the most fruitful lines of bacteriological investigation ever undertaken.[109]

In 1898 Bordet made yet another discovery of the greatest importance. It was known that the serum of normal animals tended to agglutinate the erythrocytes of foreign species and two Italian workers had just shown that the serum of a horse injected with rabbit erythrocytes acquired toxic properties for the rabbit. Immunization of animals with bacteria enhanced the power of their serum to agglutinate them. Why should not

the same process increase the agglutinating power of serum against foreign erythrocytes? Bordet promptly took a guinea-pig gave it four injections of defibrinated rabbits blood and showed that the guinea-pig's serum, ten days after the last injection, not only powerfully agglutinated rabbit erythorcytes but also caused their complete lysis. As he rapidly demonstrated, the phenomenon was exactly analogous to bacteriolysis requiring both heat-stable antibody and heat-stable complement. It was shown that antisera toxic for a variety of cells, for example spermatozoa, could be produced and the reaction appeared to be a general phenomenon by which the body dealt with any foreign cells.[110]

Haemolysis was a technically much easier phenomenon to observe and study than bacteriolysis, and its importance as a model for the experimental investigation of the mechanisms of an important immune reaction was immediately grasped by P. Ehrlich. Ehrlich was no new-comer to immunological research having been working in the same laboratory as von Behring in 1891 and, during the following seven years, had made observations of the greatest originality and importance. This is a convenient point at which to consider the immunological researches of this very great man.

Bulloch almost thirty years ago described Paul Ehrlich as the 'greatest scientific worker in medicine in the last fifty years' and it is doubtful, if he were writing today, he would feel it necessary to alter this judgement except to increase the period of years.[111] In summing up Ehrlich's work, Robert Muir wrote, 'although, as we have seen, the subjects of Ehrlich's investigations have been very varied, a unifying principle can readily be traced throughout his work. Running through it from beginning to end like a thread, as someone has said, is the question of the relation of chemical substances, natural or synthesized, to animal cells. This is seen in the domains of haematology, bacteriology, and serology, cancer research and specific chemotherapy. Originality marks his start in research, as it does all his subsequent progress – no one owed less to those who had gone before. Like Pasteur he could not be claimed by any one science; he found his field of labour for himself, and worked in it consistently and confidently. It is a long way from the staining of leucocytes to the discovery of salvarsan, just as it is from the

structure of crystals to inoculations against hydrophobia. Yet
in both cases step seems to follow step in natural sequence.'[112]

Ehrlich, of Jewish extraction, was born in Germany in 1854.
Beginning his medical studies in Breslau he followed the usual
German practice of taking courses at several other universities
passing his state medical examination in 1878, and presenting
his thesis for the M.D. degree at the University of Leipzig. This
remarkable work entitled 'Contributions to the theory and
practice of histological staining' has been said to contain the
germ of his entire life's work and attempts to elucidate the
chemical basis of the affinities of different dyes used in histology
for different tissues, to Ehrlich but examples of the specific
affinity of chemical substances for different tissue components.
He became assistant to Professor von Frerichs at the Charité
Hospital, Berlin. Pursuing his studies of the differential affinities
of various analine dyes to the cells of the blood, within three
years, he virtually founded morphological haematology, normal
and pathological. When Koch discovered the tubercle bacillus
Ehrlich, within a few weeks, described the principle of the
technique for staining these organisms which is still used today
but which, for accidental reasons, is known as Ziehl-Nielson's
method.[113] He continued to work at the Charité Hospital until
1886 producing a steady stream of highly original papers most
of which had a definite chemical slant. However his work was
interrupted when he developed tuberculosis and went to rest
in Egypt for two years. Soon after his return to Berlin, in
1890, he was offered a position in Koch's department. This was
just at the beginning of Behring's work on diphtheria which
was soon to lead to the discovery of antitoxins. In these studies
Ehrlich joined, but from his own original point of view.

From the standpoint of the history of medical microbiology
Ehrlich's work in two fields, which were largely separated
chronologically, must be considered in detail; his contributions
to immunology and to specific chemotherapy. The first of these
which will now be examined, lasted approximately from 1891
to 1903 and the latter began about 1903 and continued to his
death in 1915.

Ehrlich's contributions to the science of immunology can be
divided into five periods; his work with immunity to the vegetable poisons ricin and abrin and his connected studies on the

inheritance of immunity which occupied him roughly between 1890 and 1893. Work on the production and assay of potent diphtheria antitoxin which, of the greatest practical importance, was made, in his hands, to yield information of fundamental importance occupied Ehrlich between 1894 and 1898. His immediate appreciation of the possibilities of the immune haemolysis, described by Bordet, as a model capable of yielding information of general importance caused Ehrlich to devote the years 1899 to 1901 to an extensive study of the phenomenon. Always interested in the theoretical side and general biological significance of immunological phenomena, the first two or three years of the twentieth century saw the maturation of Ehrlich's brilliant attempt to build up a general theory explaining the numerous observations of the previous ten years. A final period up till about 1906, was devoted to fitting the rapidly accumulating new facts into his general scheme and encouraging the activities of his pupils, rather than making new observations in immunology himself.

In a very interesting letter, written by Ehrlich in 1909, he wrote that right from the beginning 'I always had the greatest interest in active therapeutics and in this combination you can find the explanation of the whole of my scientific career'. Chemotherapy had always been at the back of his mind and it was only the dramatic results of serum therapy in diphtheria, at a time when the possibilities of chemotherapy seemed remote, that diverted him into his period of immunological research. The last years of his life saw a return to chemotherapy, this time with brilliantly successful results.

Ehrlich seems to have been drawn into immunological research soon after he joined Koch's institute. Koch was at that time working on tuberculin – an extract of tubercle bacilli and its relation to disease and immunity, and Ehrlich's first paper from the department dealt with tuberculin therapy. But, with his chemical cast of mind, Ehrlich thought bacterial extracts and toxins too crude and too complex to study the fundamentals of their inter-action with the animal body. He, therefore, chose to work, not with a bacterial toxin, but with two known, highly toxic, protein, vegetable poisons – ricin and abrin. These could be extracted quite simply from the castor-oil bean and the Jequirity bean and concentrated to give a highly lethal toxin.

E

Ehrlich calculated that one gramme of pure ricin could kill one and one half million guinea-pigs.

An animal poisoned with ricin developed severe diarrhoea, prostration and haemorrhagic lesions of the intestine and died within a shorter or greater time according to the dose given. He found mice very convenient to work with and, using mice of the same weight, a particular dose of ricin gave very reproducible effects. Ehrlich found that mice could be rendered immune to a lethal dose of ricin by repeated, small, subcutaneous injections and indeed could be rendered so highly immune as to withstand 200 to 800 lethal doses at once. At first he was ignorant as to the mechanism of immunity but when, in the same institute, Behring and Kitasato showed that immunity to diphtheria and tetanus depended upon something present in the serum Ehrlich tested the power of a ricin-immune mouse's serum to neutralize that toxin and found that it would do so. Ehrlich was, therefore, in possession of an experimental model, far more convenient than animals infected with diphtheria or tetanus, on which he could study essentially similar mechanisms of immunity. He repeated his experiments with abrin instead of ricin with comparable results but he showed that the 'antiabrin', which developed in the serum of an immune animal, was specific and would not neutralize the toxic effect of ricin nor vice versa. He surmised that the serum antibody destroyed the toxins rather than rendered the animal's tissues refractory to their action.

Ehrlich decided to use his experimental model to study the inheritance of immunity and, in a series of beautiful experiments, demonstrated its mechanism in mice. He showed that mice born to a highly abrin-immune father, but normal mother, showed no immunity themselves but those born to an abrin-immune mother were themselves immune. However, the grandchildren were not immune. By allowing immune mice to suckle not only their own new-born but also the new-born of normal mice Ehrlich showed that maternal immunity was transmitted by the milk but was never to as high a degree as that of the mother. Finally, he showed that, if a nursing mouse was injected with the serum of a rabbit rendered immune to tetanus, that immunity too was transmitted to the sucklings. This work seems to have occupied Ehrlich during 1892 and was published in

three papers detailing an enormous number of most careful experiments. One other important observation on ricin immunity was made, but not until much later, when Ehrlich was deeply involved in his work on diphtheria antitoxin; this was the fact that ricin agglutinated erythrocytes *in vitro* and its activity in this respect could be inhibited by the serum of ricin-immune animals. It is a pity that Ehrlich did not notice this effect earlier because it offered an even more convenient experimental model for the study of the interaction of toxin and antibody, particularly susceptible to the quantitative and chemical techniques which Ehrlich realized were so important.

In 1896 the Prussian government established an institute in Steglitz, a suburb of Berlin, for the testing of therapeutic antisera manufactured in Germany and Ehrlich was made the director. Although it was nearly five years since Behring had introduced serum therapy for diphtheria, with undoubted benefit to patients, the basis of serum therapy was by no means wholly satisfactory and Behring's prophesy that mortality in diphtheria would eventually be reduced to insignificant proportions had not been achieved. Although the serum treatment of diphtheria had been taken up all over the world the best results were obtained on the continent of Europe and experience elsewhere had failed to demonstrate any dramatic reduction in death rate. Thus, in a series of treated and untreated cases in Connecticut, the mortality among patients receiving serum was 24 per cent as opposed to 31 per cent in the untreated group. In England results were particularly unsatisfactory, although there were many reports of individual cases and small series of cases suggesting that serum therapy was beneficial. But by 1895 serum therapy for diphtheria had to a great extent fallen into disrepute, partly because of English conservatism, partly because of a distrust of laboratory-designed therapeutic substances, following the failure of Koch's tuberculin to make good its originator's promises, and simply because British physicians had been unable to reproduce the sort of beneficial results claimed by continental doctors. *The Lancet* set up a special commission to examine the various sera available on the market for the treatment of diphtheria and soon disclosed a situation which clearly explained the variable results which had been obtained. The strengths of different serum samples

made by different manufacturers varied greatly. Thus the quantity of antiserum containing 3,000 units, a supposedly reasonable therapeutic dose, varied between 12 c.c. and 300 c.c. The continental sera were generally much more potent than the British sera, the best being made by Behring and the weakest by Burroughs Wellcome and Co. Standardization of antisera was clearly at fault.[114]

Although, in Germany, the situation with regard to the assay of diphtheria antitoxin was not nearly so unsatisfactory as in England, Ehrlich was not happy with it. He therefore began, as soon as he took up his duties at Steglitz, to work out as exact a method of assaying diphtheria antisera as possible. This routine, and at first sight humdrum task, became, in Ehrlich's hands, of the greatest interest and, within a year, he published his classical paper on 'The assay of the activity of diphtheria-curative serum and its theoretical basis'.

Diphtheria antitoxin strength had, from the first, been assayed by testing its ability to neutralize a standard toxin, so that the injection of the toxin-antitoxin mixture was harmless to a guinea-pig. But it had become apparent that the toxicity of a toxin sample grew less with age and so an antitoxin assayed against it appeared stronger than it, in fact, was. Erhlich decided that the first task of his institute must be to develop a stable standard against which new antisera could be tested, to make the actual testing procedure as accurate as possible and 'to study the complex relations which govern the neutralization of toxin and antitoxin'. He abandoned standard toxins and antitoxins kept in liquid form, with glycerine as a preservative and devised a technique for drying the substances *in vacuo* over phosphoric anhydride. Either toxin or antitoxin preserved in this way appeared stable and could have been used as a standard of reference but, for technical reasons, it was easier to prepare a dried antiserum. Ehrlich therefore prepared a quantity of standard antitoxin, a sample of which was used in the preliminary assay of the toxin, against which newly manufactured antisera were to be tested. The weight of the test guinea-pigs was kept to a standard 250 grammes and survival for four days, rather than the prevention of symptoms and local signs, was taken as the end-point of a titration. Ehrlich drew up a precise statement of the technical details for assaying diphtheria antitoxin and

these became legally binding on manufacturers in Germany, in March 1897.

But in the course of this work, 'painstaking' and 'monotonous', as Ehrlich himself described it, many anomalous results were obtained which forced Ehrlich to think deeply about the fundamentals of the subject and profoundly influenced his concepts of immunity as a whole. Ehrlich obtained a number of samples of diphtheria toxin from bacteriologists on the continent and in England and first carefully assayed their degree of toxicity against his standard antiserum. He obtained for each two threshold values (L = limes) of importance for the characterization of a particular toxin. If a standard amount of standard antitoxin was mixed with different quantities of toxin, then, 'L_o represents the quantity of toxin which is practically completely neutralized, while the other, L_+, denotes the quantity of toxin which, in spite of the antibody, such an excess of toxin is still effective that the death of the animal occurs within four days. This excess of toxin corresponds to the unit lethal dose. . . .' If the toxin was a 'Pure chemical substance' the L_+ dose minus the L_o dose must equal a 'unit lethal dose'. But, in practice, in tests involving the use of eight different samples of toxin, the difference between the L_+ dose and the L_o dose varied between less than six up to twenty-two units of toxin.

The long section of his paper entitled 'On the action of antitoxin; the theory of immunity' in which Ehrlich discussed the results do not make easy reading. Ehrlich regarded it as certain that a molecule of toxin combines with a definite and unalterable quantity of antibody', fitting together by complementary groups of atoms 'as a key does a lock'. On general grounds he felt that, in antibody production, one had to deal with an 'enhancement of a normal cell function' rather than that cells developed the ability to produce entirely new kinds of molecules. It was demonstrable that, for example, tetanus toxin injected into an animal became bound to the cells of the central nervous system and, Ehrlich postulated, that it did this by becoming attached to certain chemical 'side-chains' on the cell protoplasm and thus the physiological function of the side-chains became blocked. In response to this, the cell produced fresh side-chains and produced them to excess, antibodies being no more than 'side-chains of the cell protoplasm which have been produced

in excess and, therefore, thrust off'. This was Ehrlich's theoretical background at the beginning of his work on the assay of diphtheria antitoxin and into which he attempted to fit his findings.

He had appreciated, since 1893, that the neutralizing capacity of a toxin and its absolute toxicity were not inseparably linked. He had observed that if tetanus toxin was mixed with 'carbon sulphide' it lost its toxicity so that even 1 c.c. could be injected into a mouse without harm but such mice were found subsequently to have developed immunity to tetanus toxin. He had also shown that such modified toxin could combine with antitoxin and coined the name 'toxoids' to designate them. He noted that diphtheria toxins, after an interval of some months, might lose a substantial part of their toxic power yet still be able to combine with the same quantity of antitoxin. The variable differences between the L_o and the L_+ dose of different samples of toxin could be explained on the basis that each contained, in addition to toxin molecules, variable amounts of toxoid. Ehrlich elaborated this concept by postulating that toxoids varied in their affinity for antibody molecules having either the same affinity (syntoxoids), greater affinity (prototoxoids) or less affinity (epitoxoids) and it was the presence of these last which explained the unexpected difference between the L_o and L_+ dose of toxin.

In order to explain differences observed with different batches of fresh toxins Ehrlich postulated the existence of 'toxones', substances elaborated by the diphtheria bacillus at the same time as the toxin, but which had less affinity for antitoxin than either toxin or toxoid. He suggested that although the toxones were not responsible for the killing effect of the diphtheria culture filtrate, they probably were responsible for the local reaction and for the slowly developing paralysis. By mixing a constant amount of toxic filtrate with differing fractions of its neutralizing dose of antitoxin and measuring the degree of toxicity of the mixture Ehrlich thought he obtained a picture of the composition of a toxic filtrate which he called the 'toxin-spectrum'.

The action of toxins and antitoxins formed the subject of research for a large and active group of workers all over the world who constantly threw up new, sometimes contradictory,

and always confusing, observations. These Ehrlich selected and sifted and combined with findings of his own or his pupils in the constant endeavour to gain a clear concept of action of bacterial toxins in the living body. A number of experiments suggested that an anti-toxin molecule contained two distinct groups, one which attached the molecule to the cell (haptophore group) and another distinct group which exerted the toxic effect (toxophore group). A neat experiment by one of his pupils, Dr Morgenroth, was one of several supporting this idea. Morgenroth injected frogs with a lethal dose of tetanus toxin but showed that if the frogs were kept cool they did not suffer tetanic spasms – they did so when allowed to become warm. The tetanus toxin had, however, clearly become attached to the cells of the nervous system while the frogs were cool since full doses of antitoxin immediately prior to warming them did not prevent tetanic spasms. The combining and toxic fractions of the toxin molecule must, therefore, be distinct.

In 1899 Ehrlich resigned his appointment in Berlin and went to Frankfurt-on-Main to direct the newly-founded Royal Prussian Institute for experimental therapy. Here he remained for the rest of his life.[115]

Just before Ehrlich's departure from Berlin, Bordet had published his account of immune haemolysis which Ehrlich immediately recognized as a system particularly suited to the *in vitro* study of the action of antigen and antibody by the quantitative methods which he felt would lead to a proper understanding of the subject. With his assistant, J. Morgenroth, Ehrlich immediately began his investigation of immune haemolysis and, over the next six years, published a series of classic papers on the subject which must now be considered. Before so doing a few brief biographical details of his associate in this work, Julius Morgenroth, will not be out of place. Morgenroth was 27 years old when he began his work with Ehrlich at Steglitz and moved with him to Frankfurt. He had been born in Bauburg in 1871 and studied with Weigert in Frankfurt. After working with Ehrlich for some years he went to the Berlin Pathological Institute and later to the chemotherapeutic department of the Koch Institute. He died, in 1924, of pernicious anaemia.

Ehrlich pointed out that it was the obvious dose analogy

between the phenomenon of haemolysis and bacteriolysis which gave the former its considerable theoretical significance. Enthusiastic as he was at this date for his newly-propounded 'side-chain theory', Ehrlich was particularly concerned to fit the new observations on haemolysis into his general scheme. To do this was reasonable and almost certain to be rewarding, but, if fault can be found with this great genius, it lay in an excessive zeal to fit new observations into his theoretical framework rather than make any radical change in his hypothesis. The science of immunology was so new and fresh observations, on many aspects of the subject, were pouring in all the time so that, valuable as a theoretical scheme was as a guide to further experiment, Ehrlich to some extent hampered his own development by too slavish adherence to his very remarkable 'side-chain' theory. His first contribution to the subject of immune-haemolysis was entitled 'contributions to the theory of lysin action'. The paper did, however, also report a clear analysis of the haemolytic phenomenon, based on his own experiments, using the serum of a goat which had been immunized with sheeps serum, but serum from which the red cells had not been carefully removed and which, therefore, was found to be haemolytic. He happened to have a large stock of it. But straight away the fortuitous choice of species of animals led to an anomalous result; his haemolytic serum showed no power to agglutinate red cells, as had Bordet's guinea-pig anti-rabbit red cell sera. Ehrlich, therefore, concluded that haemolytic and agglutinating antibodies were quite distinct. He showed, by mixing heated antiserum and red cells, followed by centrifugation that the antibody became attached to the red cells and, therefore, must have a haptophore group and by similar experiments demonstrated that the necessary haemolytic substance (complement) which he called 'addiment' had no such group. He postulated that the antibody molecule must have two haptophore groups, one to combine with the cell and the other to link up with complement. Complement he regarded as a ferment and it was the function of antibody, with its second haptophore group, to concentrate the ferment on the surface of the red cell. This Ehrlich likened to the insectivorous plant, Drosera, whose tentacles first grasp an insect and then secrete a digestive ferment. The whole phenomenon was but an example of 'a process of

normal cell life'. The side-chains normally anchored large molecules to the cell surface and at the same time brought the complement, ferment molecule to bear for its digestion.

Ehrlich and Morganroth continued their researches along two lines; they manufactured two new samples of goat-anti-sheep red cell serum and analysed the limited lytic power of normal blood serum for the red cells of foreign species of animal. Their goat immune sera led to a further elaboration of the factors required for red cell lysis, since one of the antisera required more prolonged heating to destroy its haemolytic effect – these were, therefore, two different sorts of complement. Yet the probable explanation of this observed effect lay in the relative proportions of antibody and complement in the two antisera. Ehrlich was ever inclined to explain a new phenomenon by postulating a new substance in the serum, rather than an existing substance acting in a different way. Analysis of the lysis produced by normal, non-immune, serum they showed to be essentially similar requiring a heat-stable component, which Ehrlich distinguished from the immune-body by the name of 'interbody' and heat-labile complement. They showed that complement was relatively non-specific, in that the complement of one animal could activate the immune-body produced in a different species. But this non-specificity was not absolute since, for example, eel 'interbody' could not be activated with the sera of the usual animals available.

In 1900 Ehrlich was invited to give the Croonian lectures before the Royal Society. He spoke 'on immunity with special reference to cell life' and summarized his side-chain theory, the history of the development of which has just been traced.[116] The next four years were ones of intense activity in immunology. His main efforts went into four more subsequent papers, 'on haemolysins', but he wrote several more substantial papers on antitoxin, the nature of complement and red cell receptors. He also wrote several polemical papers, arguing against objections to his side-chain theory, raised by Max Gruber and Svante Arrhenius. The number of workers in the new field of immunology was already considerable partly because the requirements for the study of the subject, two or three species of experimental animal, syringes, pipettes, test-tubes and saline solution, were very simple. The result was a stream of observations on very diverse

aspects of the immune reaction. Ehrlich himself contributed a vast number of new observations, in many respects of greater value than most, because his experiments were guided by his side-chain theory. None the less he felt it encumbent upon himself to fit all the reported observations of other workers into his general theory and, although to a surprising degree successful, the task became progressively more difficult.

Among the particular problems Ehrlich himself investigated was the possibility that the body might be able to make an immune response to some of its own components, especially in pathological situations. He showed that one goat could indeed make an antibody against the red cells of another goat but considered that, in general, the body did not react against its own tissues; a situation he designated 'horror autotoxicus'. But that the body might so react in pathological circumstances he was convinced, postulating a control mechanism remarkably similar to the 'homeostatic mechanism' of Sir Macfarlane Burnet. Referring to the possibility of an auto-immune disease, Ehrlich remarked that usually this would not occur, but, 'only when the internal regulating contrivances are no longer intact can great dangers arise. In the explanation of many disease-phenomena, it will in the future be necessary to consider the possible failure of the internal regulation, as well as the action of directly injurious exogenous or endogenous substances'.

In 1904 Ehrlich was invited to give the Herter lectures at Johns Hopkins University. He took as the titles for his lectures the 'Mutual relations between toxin and antitoxin; physical chemistry versus biology in the doctrines of immunity and cytotoxins and cytotoxic immunity'. These lectures may be taken as Ehrlich's matured views on the fundamentals of immunological science, for his main attentions were soon diverted to the fields of cancer research and chemotherapy. However, in his Herter lectures, Ehrlich added little to his contribution to immunology. He was mainly concerned to fit into his side-chain theory certain new observations by other workers; for example, Park's demonstration that toxin neutralized with antitoxin could still produce further antitoxin if injected into an animal. Ehrlich was compelled to postulate yet another toxin fraction, the 'ultra-toxoid'. He also devoted much time to answering the many criticisms that had been directed against his side-chain

theory. His most important critics were Jules Bordet, Svante Arrhenius and Max Gruber, with their respective pupils, and at this point, their objections and counter-suggestions must be examined.

The contributions of Bordet to the science of immunology rival in importance those of Ehrlich. Both scientists worked with almost exactly the same tools, biological systems and techniques and sincerely admired each other's work, yet differed materially in their fundamental views on the phenomena of immunity. These differences can best be appreciated by quoting fairly extensively from a general résumé of immunity published by Bordet in 1909.[117]

Bordet objected to Ehrlich's side-chain theory, writing that, 'Its principal fault to my thinking is that it is not, strictly speaking, a theory, but rather an assertion of a certain number of undemonstrated facts.' For example, Bordet could not accept that antibodies were cell-receptors discharged into the blood; they might be, but one could put forward other hypotheses which fitted the known facts just as well. He felt that Ehrlich's theory had 'exercised a perturbing influence on the progress of knowledge, and has really hindered the free development of investigation. In offering explanations which seem definitive, and schemata which satisfy the experimenter and appease has curiosity, Ehrlich's theory has come to make certain problems, which have scarcely been touched upon, regarded as worked out'. Bordet characterized his own method of working by saying', 'I have yielded as little as possible to the inspiration of theory; and for this reason, moreover, no general conception of obscure questions will be found in the present article. Like every other observer, I have offered certain hypotheses, but they scarcely merit this name, for they are so little removed from the facts observed; they are rather a transcription of impressions gathered from the results of laboratory experimentation. At the risk of being considered by some readers as not possessing a sufficiently generalizing mind, I must admit that I have been led to make my most important discoveries by yielding tractably to the impulse of facts, by letting myself be moved by my data without attempting to discipline them or subject them to systematic ideas of my own.'

Bordet's positive contributions to immunological theory were

to 'deny definitely any chemical character to the union of anti-body and antigen' which union he thought more akin to 'what is called molecular adhesion or contact affinity, in other words should be classed in the category of absorption phenomena'. He showed experimentally that there were close analogies between the dying of a piece of filter paper by a weak solution of an analine dye and the ability of antigen and antibody to combine in multiple proportions and offered a good explanation for the Danyz phenomenon. Bordet conceived of antigen-antibody union as a two-stage phenomenon, the first stage, actual union, was undoubtedly specific but the nature of any subsequent visible reaction, such as agglutination or lysis by complement, relatively non-specific and dependent upon the exact conditions in the reacting mixture. He himself had shown, for example, that whether or not certain bacteria mixed with specific antibody agglutinated depended on the electrolyte concentration of the suspending fluid. He did not regard the various phenomena of antigen-antibody reaction as due to different sorts of antibody, as did Ehrlich, but as varied manifestations of a single antibody.

Bordet differed from Ehrlich with regard to the nature of complement and its mode of union with anti-antibody complexes. Bordet insisted that the body did not contain a whole series of different complements but that, 'on the contrary it is the same weapon in each instance, a single alexin (complement) which reacts now against one and now against another bacterium, owing to the specificity of the sensitizer'. This was a most important point and arose from some of Bordet's most significant work in immunology. In 1901 he published a paper with O. Gengou describing a new method for detecting antibodies to a wide variety of bacteria, based on his studies of immune haemolysis and the bacteriolysis of cholera vibrios. They pointed out that 'to demonstrate the existence of a sensitizer (antibody) in an antimicrobial serum we may make use of its property of causing the bacterium it affects to absorb alexin (complement)' and went on to describe the principle of the complement fixation test, in which the bacterial antigen, complement and the serum suspected of containing antibody are mixed and subsequently the presence of free complement tested for by adding red cells sensitized with antibody. They showed that all

the organisms tested were capable of absorbing complement when sensitized with specific antibody and concluded that 'specific sensitizers are formed in vaccinated animals as a general rule'. The validity of the complement fixation test depended upon the well-founded assumption that the complement which caused bacteriolysis was the same as that causing haemolysis. Bordet by no means regarded the complement found in all species of animal sera as identical but, for the purpose of function, they behaved in very similar ways and there was no need to postulate a multiplicity of complements as Ehrlich had done.

Criticisms of Ehrlich's theories along lines rather different from Bordet came from two Scandinavian workers, Thorwald Madson and Svante Arrhenius. Both worked for a time in Ehrlich's laboratory and later transferred to the State Serum Institute, in Copenhagen, of which Madson became the director in 1902. Their views on immunity were presented in some detail by Arrhenius in a series of lectures, for which he coined the title 'Immunochemistry', given at Berkeley, California, in 1904, and subsequently published in book form.[118]

Arrhenius set out to investigate the relations between toxins and antitoxins according to the general principles of physical chemistry. Drawing on a large number of observations of different immune phenomena, in the literature as well as his own experiments, he had no difficulty in drawing attention to many facts which seemed inconsistent with Ehrlich's notion of antibody-antigen combination as being one of firm chemical union; various observations suggesting that antibody and antigen could combine together in different proportions, that their union was often readily reversible and the slow velocity of their reaction together. Arrhenius, working with Ehrlich's colleague Morgenroth, demonstrated that a constant quantity of erythrocytes absorbed increasing amounts of antibody as the concentration of antibody was increased in a manner consistent with the application of the law of mass action, such as operates when weak acids and bases are mixed. Arrhenius and Madson had considerable success in applying this law to the large number of quantitative observations on antibody-antigen reactions to be found in the literature. They, justifiably, ridiculed Ehrlich's school of thought which had to invoke the presence of a new substance to explain every newly discovered phenomenon but

could not agree with Bordet's idea that the reactions of antigens and antibodies had much in common with colloid chemistry. Their objections to Bordet's point of view were much less well founded and hinged on rather small points, considering that most of the observations they cited against Ehrlich's theory could equally well be consistent with the analogy with colloid chemistry.

6 The Practical Application of Immunology to Medicine

The birth of modern immunology can be dated to the discoveries of Pasteur relating to the artificial induction of immunity to anthrax and chicken cholera about 1881. So far we have followed the development of the science through its great triumphs in practical medicine, rabies vaccination and the serum therapy of diphtheria, and considered the history of our knowledge of the underlying mechanisms of immunity taking the story down to about the middle of the first decade of the twentieth century. But for all the great interest of the academic study of the interaction of antigen and antibodies the driving force behind the whole effort was the severely practical aim of preventing and treating human infectious disease. We must now turn to see what progress had been made in this direction, apart from the great successes of rabies prophylaxis and diphtheria therapy, during the quarter of a century following Pasteur's demonstration of the possibility of immunization as a means of combating infectious disease.

Pasteur's work on anthrax could not immediately be applied to human disease for, by 1881, none of the causative organisms of any of the major human bacterial diseases had been isolated. However, during the next five or six years a number of human bacterial pathogens, the causative organisms of important epidemic diseases, were discovered – the tubercle bacillus, the typhoid bacillus, the cholera vibrio, the micrococcus of Malta fever and the pneumococcus, for example.

The first attempt to extend the principle of prophylactic immunization to man was made in the case of cholera and credit for this attempt must go to the Spanish bacteriologist, J. Ferran. Ferran was born in 1852 and graduated in medicine

at Barcelona. From his student days he was an enthusiastic bacteriologist and, in 1886, became the director of the municipal laboratory at Barcelona. He made no important conritbutions to science but was an early and enthusiastic advocate and practitioner of prophylactic vaccination against a wide variety of diseases. He died in 1929.[119]

In the spring of 1885 cases of gastroenteritis became numerous in Jativa and Valencia and on 12 April, the authorities admitted that there was an outbreak of asiatic cholera. Dr J. Farran was on the spot immediately investigating the disease and proposed mass immunization on the principle which Pasteur had successfully applied to the immunization of sheep against anthrax. Ferran immediately confirmed that the organism described by Koch was found in the Spanish cases of cholera. (This in itself shows that Ferran was a competent bacteriologist; a British commission of experts sent to India failed completely to confirm Koch's findings.) He distinguished the cholera vibrio from Finkler's non-pathogenic vibrio and showed that injections of the former were pathogenic for guinea-pigs. However, he noted that guinea-pigs which happened to survive inoculation were immune to subsequent challenge with a fatal dose. In only one respect was Ferran in error – he described various non-existent stages in the life-cycle of the cholera vibrio, including spores. Following their guinea-pig experiments Ferran and his colleagues submitted themselves to subcutaneous inoculations of living cholera vibrios. The first injection produced local inflammation, general malaise and a mild fever but the second injection caused only local irritation. These experimental results, few in number and hastily carried out, together with the analogy with Pasteur's work, formed the basis upon which mass inoculations were carried out. Ferran soon produced statistics purporting to demonstrate the efficacy of his inoculations. Thus, in Alcira, among 5,432 inoculated persons there were no deaths from cholera but among the remaining 10,500 of the population there were thirty-four deaths from cholera.[120] A Spanish Royal Commission fully endorsed Ferran's work (for he had been subjected to criticism, particularly by foreigners) and he was given permission to carry on with his inoculations. Meanwhile, the cholera spread rapidly and, when Ferran began operations in Valencia itself, hundreds flocked to him for inoculation even

at the cost of about ten shillings per injection. The French and Belgian governments sent bacteriologists to study cholera on the spot and, in particular, to assess Ferran's preventive inoculations. Ferran's techniques were crude and there were doubts about the purity of his cultures but it is fair to point out that Dr Van Emmergen, the Belgian, considered Ferran's methods crude but adequate. Ferran himself was described as 'a most approved disciple and enthusiastic follower of the school of Pasteur'. But Van Emmergen pointed out that Ferran's Alcira statistics were utterly fallacious and no account had been taken of the variability of risk of infection between the inoculated and un-inoculated. The former group contained large numbers of well-to-do who were less exposed to infection. Moreover, Ferran claimed that his vaccine consisted of attenuated organisms but declined to say how he produced his vaccine or supply other workers with samples.

Faith in Ferran, even among Spaniards, was short lived. In June he was prohibited from making more inoculations and deprived of his cultures and syringes until more was known about the nature of his vaccine. Meanwhile Ferran claimed the prize of 100,000 francs which had been offered by the Academy of Sciences of Paris for an effectual remedy against cholera. A point reasonably enough raised by those sceptical of Ferran's immunization method was the fact that there was no evidence that one attack of true cholera conferred any subsequent immunity on those who recovered; indeed such evidence as there was suggested that the contrary was true. Certainly there is little to suggest that Ferran's inoculations did anything to stop the epidemic which spread from province to province and, at its height, was killing 2,000 persons a day. The epidemic died away with the coming of autumn. Another Spanish commission which investigated Ferran's claims concluded that there was no evidence that his vaccine actually was a culture of attenuated cholera vibrios and certainly no proof that it afforded any immunity to the disease.[121] Ferran's work on cholera has been described in some detail since it represents the first attempt to extend the Pasteurian principle of prophylactic immunization from animals to man. Ferran deserved credit for the attempt and also for the choice of disease, for an effective procedure would be more obviously beneficial in a fulminating epidemic

of cholera than in a disease the epidemic spread of which was less dramatic. It would be unfair to blame him because his methods were crude, for keeping the nature of his vaccine secret (did not Koch, later, do the same with tuberculin?) or for the fact that he had no idea how to organize a trial to see whether or not his immunization was effective. In this last respect he was at one with most of the famous 'immunizators' who were to follow him.

The first vaccinations for bacterial disease in man the utility of which seemed reasonably certain were made by W. M. W. Haffkine in 1893.[122] He, like Ferran, worked with cholera. Haffkine was born in 1860 in Russia, the son of a poor Jewish schoolmaster. He studied science at the University of Odessa, in particular zoology under Metchnikoff who was, at that time, professor there. Metchnikoff greatly influenced Haffkine and also saved him from imprisonment for his student political activities. After graduation in 1883 Haffkine continued to work as a zoologist in Odessa for five years until increasing repression and persecution of Jews led him to emigrate in 1888. For a year he worked as an assistant in the physiology department of the University of Geneva but then sought out his old teacher, Metchnikoff, who had settled in the Pasteur Institute in Paris. The only appointment which could be found for Haffkine at the Pasteur Institute was that of librarian but, happily, his duties allowed him time for laboratory research which he carried out in Metchnikoff's department. His first publication after joining the Institute reflected Metchnikoff's interests, being on the infectious diseases of the protozoon Paramecium. But he soon struck out on his own and began investigations of cholera in laboratory animals about 1891. Haffkine set about his attempt at prophylactic immunization in cholera in a manner analogous to Pasteur's method with rabies. He found that by repeated passage through the peritoneal cavities of guinea-pigs the virulence of cholera vibrios for that animal could be enhanced about twenty times, but no more; it had reached the stage of 'fixed virus'. Such strains of enhanced virulence would also kill rabbits and pigeons, and in addition, if injected subcutaneously caused a severe local reaction with skin necrosis, whereas ordinary strains would not. Haffkine also found that cholera vibrios grown at 39° C. in broth through which air was bubbled became

less virulent and would not produce skin necrosis following subcutaneous injection. He had thus provided himself with two strains of cholera vibrio of differing virulence for guinea pigs and he showed that, if an animal was first inoculated with the less virulent strain, subsequent subcutaneous inoculation of the virulent strain did not cause skin necrosis and that a guinea-pig which had received both strains was completely protected by whatever route it was challenged, either with virulent laboratory strains or 'wild' strains of cholera vibrios obtained from Madras, Calcutta and Saigon.

On 18 July 1892, Haffkine inoculated himself with his attenuated strain of cholera vibrio producing a severe local reaction and fever. Six days later he tried his virulent strain with the same result. He persuaded three Russian friends to undergo inoculation; all suffered from local reaction and fever but survived. Haffkine, therefore, considered that his immunization was safe and added, 'I express the hope that six days after vaccination man will acquire immunity against infection with cholera.'[123] It happened that the British Ambassador in Paris was Lord Dufferin, a former Viceroy of India, and it was through his good offices that Haffkine was given permission ro go to India to try out his vaccine in the field. In November 1892 he went to England to make arrangements and during the course of his visit went to the Army Medical College at Netley. This visit had the important effect of interesting A. E. Wright, the professor of pathology, in the subject of prophylactic immunization. Haffkine began work in India in April 1893 and spent the next twenty-nine months vaccinating over 40,000 persons. The people vaccinated varied very much in their social circumstances and in their risk of exposure to cholera. No pressure was put upon anyone to be inoculated nor could it be said that a trial of the vaccine in a formal way was made. However, Haffkine kept careful records and made every attempt to compare the incidence of cholera in vaccinated and unvaccinated people subject to the same risk of infection. Administrative and technical difficulties rendered many of his groups unsuitable for the evaluation of the vaccine and Haffkine himself, by experimenting with various dosages, introduced yet other variables. But, although the apparent effectiveness of vaccination was very variable, his better experiments strongly

suggested that it was a valuable procedure. Thus in one group of people, living in huts around the tanks in Calcutta, who had received two medium doses of vaccine and who were observed from 11 to 459 days thereafter it was found that cases of cholera occurred in 26 houses. In those houses there were 263 un-inoculated persons of whom 34 died of cholera and 137 inoculated persons of whom only 1 died. Haffkine considered his results to be very promising and discussing them with Koch the latter expressed himself convinced.[124]

In the autumn of 1895 Haffkine took leave in Europe but was keen to return to India to continue his work on cholera prophylaxis and this he did in March 1896. But soon after his return he was asked by the Indian government to go to Bombay to investigate an outbreak of bubonic plague. He abandoned his cholera work and set up a small laboratory in Grant Medical College. The causative organism of plague had been discovered, in 1894, in Hong Kong, by Kitasato and, indeed, it was probably the plague that had spread by ship from Hong Kong which was causing the epidemic in Bombay. Haffkine rapidly confirmed Kitasato's work and stressed the necessity for accurately identifying the plague bacillus. In this connection he described two cultural characteristics; stalactite growth in broth and certain involution forms on agar. Naturally he turned at once to the possibility of prophylactic immunization. Haffkine had a sound grasp of the principles underlying immunity in so far as they were understood at the time and his attempt to prepare a plague vaccine were by no means crude. He realized that immunity might depend on the production of bactericidal antibodies or antitoxic antibodies or perhaps both and that, therefore, a vaccine ought to contain a high concentration of both organisms and their metabolic products. He devised a broth medium, the surface of which was covered with butter fat, on the undersurface of which the plague bacilli grew as stalactites and could periodically be shaken off into the broth and a fresh crop grown. He thus built up a high concentration of organisms and their metabolic products. He made no attempt to attenuate the plague bacillus but prepared his vaccine by killing them with heat at 70° C. The vaccine was first tried out, in January 1897, on the prisoners in a Bombay jail where plague had already broken out. Vaccination was offered on a

voluntary basis and happily produced two roughly comparable groups. In the following months there were 12 cases of plague with 6 deaths among 173 non-vaccinated prisoners and but 2 cases with no deaths among 148 vaccinated prisoners. In addition, during the first five months of 1897 over 11,000 members of the general population in Bombay were immunized and it was estimated that they suffered only about one-twentieth of the incidence of plague in the immunized population.[125]

Haffkine continued to run and enlarge the Plague Research Laboratory in Bombay until, reaching the age of 54, he retired from India in 1914. One unhappy incident marred his otherwise useful and distinguished career; the Malkowal disaster. At this village, in 1902, nineteen persons inoculated with his vaccine died of tetanus. An official inquiry ultimately completely cleared Haffkine of responsibility; it seemed the vaccine had become contaminated by gross carelessness at the time of administration, not in the laboratory. None the less, during the inquiry, which lasted several years, Haffkine was humiliated and suspended from his post as director of the plague laboratory in a quite unjustified manner and the event left its mark upon him. In the end the government to some extent made up for its treatment of Haffkine – in 1925 the plague laboratory was renamed the Haffkine Institute. Haffkine was a rather solitary individual who never married. But he seems to have been popular with all those who worked with him. It is true that in estimating his place in the history of immunology it must be remembered that he introduced no new principle nor added significantly to our understanding of the phenomena of immunity. But, to have been the first person to apply prophylactic immunization to man, to have grappled with the technical and administrative difficulties involved and pioneered the general procedure which has made possible the eradication of such diseases as diphtheria, poliomyelitis, tetanus and many other infections entitles him to a distinguished place in the history of medicine.

This is a convenient point to consider the contribution of one of the most influential workers on immunity during the decades on either side of the year 1900 – Sir Almroth Wright. A. E. Wright was the son of a clergyman living in Ireland and qualified in medicine at Dublin, in 1883, when 22 years of age. Being determined on a career in research he immediately

obtained a studentship which enabled him to work at Leipzig for a year. Here he met another British post-graduate student, L. Wooldridge, who was greatly to influence the line of work Wright did for a number of years. During the following eight years Wright supported himself in a variety of ways, including taking a studentship in law and working as a clerk in the Admiralty, all the time enlarging his research experience. His first scientific papers were not published until 1891, when he was 30 years old, and stemmed directly from the work of his friend Wooldridge whose main interest was in the mechanisms of blood-clotting. But Wooldridge also had some interest in the problems of immunity and had described a method which, he claimed, would immunize rabbits against anthrax, consisting of injections of what he called 'tissue-fibrinogen', actually extracts of testis or thymus gland. Wright's study of this material was mainly chemical. He showed that its chief constituent was nucleic acid and prepared material with similar properties from yeast. He was unable to demonstrate any immunity to anthrax in animals inoculated with his tissue extract but, in a very small-scale experiment, in which one of his two control rabbits survived challenge with anthrax bacilli, he concluded that injections of his yeast extract did prolong the survival time of anthrax infected rabbits.[126]

Wright's interests at this time were not immunological but, in January 1893, he was visited at Netley by Haffkine who demonstrated his technique for preparing attenuated strains of *Vibrio cholerae* and showed, in a small experiment, that vaccination of guinea-pigs enabled them to withstand up to five times the dose of virulent organisms which killed control animals. Wright was completely convinced of the validity of Haffkine's results and described the technique in detail in the *British Medical Journal*. It is worth noting that although Haffkine relied principally on the use of live, attenuated cholera organisms as a vaccine he also began his course of injections, at least in his Netley experiment, with a suspension of phenol-killed organisms.[127]

For the next three years Wright was busy with his work on blood coagulation and related matters and did no work on immunity. In 1896 he reported the first injection of heat-killed typhoid bacilli into man, but merely because he thought that

calcium chloride might reduce the local reaction to various oedematous and haemorrhagic reactions in the skin and knew that Haffkine's cholera vaccine produced such reactions. He had no intention to try to produce immunity and all he demonstrated was that, although dead typhoid bacilli caused local and general reaction, it was apparently a safe procedure.[128] Wright sent a reprint of this paper to Pfeiffer who also inoculated a volunteer with dead typhoid bacilli but, the agglutination reaction having just been described by Gruber and Durham, was interested to see whether or not his inoculated volunteer had developed serum agglutinins. He showed that he had and, in conversation, told Wright about it. Wright, who was at the time adapting the agglutination reaction for the diagnosis of Malta fever and experimenting with the immunization of monkeys against that disease, saw the significance of Pfeiffer's finding for typhoid immunization. He promptly inoculated eleven laboratory workers with heat-killed typhoid bacilli, producing a striking local and general reaction, and three of his volunteers 'looked somewhat shaken in health for some three weeks after'. They developed serum agglutinins against the typhoid bacillus and one such volunteer was inoculated with living typhoid bacilli without untoward effect. Taking all that was known about the development of agglutinins in convalescence from typhoid fever, the undoubted fact that guinea-pigs could be immunized against cholera vibrios and his own few experiments, Wright leapt to the conclusion 'that the sedimentary power of the blood is a trustworthy criterion of the immunity of the person who furnishes it' and that 'the possession of a sedimentary power connotes also the possession of a certain measure of bacteria-proofness'. He glossed over the fact that persons who died of typhoid also had agglutinins in the blood, a fact of which he was well aware.[129]

On this rather slender evidence Wright was convinced, and of course he was correct, that vaccination with dead typhoid bacilli provided at least some immunity against the disease. The next few years were devoted to getting groups of people, mostly British Army personnel, vaccinated whenever possible and comparing the subsequent incidence of typhoid fever among them with as nearly comparable as possible controls.

His first data came from the staff of a mental hospital at which

there was an outbreak of typhoid and where, out of a staff of about 200, twelve had already contracted the disease. Eighty-four of the staff volunteered to be vaccinated whilst 116 declined. The subsequent incidence of typhoid in the two groups was nil and four respectively.

About this time an incident occurred which might well have led Wright to abandon his poorly substantiated view on the value of prophylactic immunization. He was still experimenting with brucellosis and attempting to immunize monkeys with dead cultures, when he resolved to try the procedure on himself. In the early part of 1898 he was inoculated with three doses of dead brucella organisms, spaced over four weeks, and, two weeks later, was inoculated with a suspension of live brucella. Wright came down with a severe attack of brucellosis which made him miserably ill throughout the spring and summer of 1898.[130]

Happily he was undeterred as regards typhoid immunization and patiently accumulated data about the incidence of typhoid in inoculated and uninoculated troops in India, Cyprus, Egypt and South Africa. He was well-aware that the records on which its statistics were based were, for various reasons, not wholly reliable but concluded that there was evidence that inoculation reduced the incidence and mortality from typhoid. Meanwhile Wright continued to study typhoid immunization experiment-ally, at Netley, using surgeons-on-probation as guinea-pigs, and, in 1901, published a long paper 'on the change effected by antityphoid inoculation in the bactericidal power of the blood with remarks on the probable significance of these changes'. The work here reported is of importance because of its effect on Wright's views on immunization in general. In a number of different individuals he determined the capacity of their serum, before and after immunization, to sterilize an aliquot of a culture of typhoid bacilli. Typically he found that before immunization 1.0 ml. of serum would sterlize a 1 in 10,000 dilution of typhoid culture but, the day after an inoculation of dead typhoid bacilli, would sterilize a 1 in 50 dilution and, two months later, a 1 in 5 dilution. But, in certain individuals, before the enhancement of the blood's bactericidal power there occurred a 'negative phase', which might last up to three weeks, during which the bactericidal power of the blood was less than

it had been before inoculation, although subsequently, it became more bactericidal. Thus, in one individual, the pre-inoculation serum sterilized a 1 in 100 dilution of typhoid bacilli yet, eight days after inoculation, failed to sterilize a 1 in 10,000 dilution.

As a result of much work Wright concluded that it was only doses of vaccine sufficiently large to cause a constitutional upset which might result in a 'negative phase' and that, probably, to produce such a 'negative phase', in a person exposed to natural typhoid fever, would be dangerous. However, small doses of vaccine did not give a 'negative phase' and might induce increased bactericidal power within twenty-four hours. Wright considered that the results he obtained with typhoid inoculations were exactly analagous to those obtained by such workers as Ehrlich with toxins and antitoxins and therefore 'entitled to rank as a general principle of immunization'. He failed to notice certain differences, for example, the rapid effect, one way or the other, of his inoculations and, in fact, Wright was probably dealing with a more complicated situation involving non-specific depression and enhancement of reticuloendothelial system activity by endotoxin as well as the production of antibodies.[131]

At this point it is necessary to say something about the history of the application of statistical techniques to the problems of immunology. Statistical considerations had arisen, to some extent, in connection with earlier immunological forms of prevention and treatment of diseases such as smallpox and diphtheria but they first came into prominence in relation to antityphoid vaccination. It is difficult to exaggerate the importance of a sound statistical approach in the assessment of immunological procedures, whether prophylactic or therapeutic but, unhappily, at the beginning of the twentieth century the need to consider this aspect of the subject was not appreciated by the medical profession. The confused state of immunology which W. W. C. Topley was to lament in the 1930s 'with its mixture of established fact, half-knowledge, hopeful guessings and frank bewilderment' might have been avoided had bacteriologists heeded the suggestions of non-medical statisticians.

At the end of the South African War, during which there had been a very heavy mortality among British troops from typhoid

fever, and some soldiers had been vaccinated against typhoid, the army authorities had to decide whether or not to continue with antityphoid inoculation and make it compulsory or abandon the procedure. A number of committees sat to consider the evidence on the subject and the details of their deliberations need not be followed. However, in 1904, the opinion of an eminent statistician, Professor Karl Pearson, was sought. Pearson, at the time of his controversy with Wright, was 47 years old and Goldsmith professor of applied mathematics and mechanics in University College, London. His main interests lay in statistics, eugenics and the application of the mathematical theory of probability to biological data. In 1901 he had founded the journal *Biometrica* for the publication of studies of this kind. There is no doubt that he was well qualified to advise on the statistical aspects of antityphoid inoculation. Pearson published the result of his statistical analysis of Wright's antityphoid statistics in the *British Medical Journal*. There ensued an acrimonious controversy between Pearson and Wright which may be followed in the second volume of that journal for 1904. This controversy was of great significance for the development of immunology. Pearson's point of view, which embraced more than the immediate question of antityphoid inoculation, may be summarized as follows: He held that there was 'a crying need for a more exact treatment of statistics in medical science' and, overcrowded though the medical curriculum might be, advocated some instruction in medical statistics for students. Failing that, he thought that at least individual doctors, working on problems with a statistical side, might seek instruction or consult a statistician. He suggested that all published tables of experimental data should contain 'a mathematical expression for the effect exerted by the operation of chance'. Examining the antityphoid inoculation statistics he applied a technique of the 'coefficient of correlation', in which a complete absence of correlation between two measurements would have a coefficient of nought and complete correlation a coefficient of one. Measurements which were undoubtedly closely correlated gave high figures, say above 0.75. Pearson found that the evidence for a correlation between antityphoid inoculation and low incidence of the disease and low mortality were 'sensible', that is, were in the right direction but the coefficient of correlation was of a

low order; about 0.20 and appeared 'to fall into that range of intensity which would justify the suspension of the operation as a routine method. . . . The differences on which stress is laid by Dr Wright in his book are largely of the order of the probable errors of random sampling'. However, he suggested that further careful assessment of antityphoid vaccination should be undertaken.

Wright, sincerely convinced of the value of antityphoid vaccination, was incensed, and, of course, in this particular instance his view has been shown to be correct. His attitude is apparent from his replies to Pearson and from his discussion of statistics in his book on antityphoid inoculation. Wright was not wholly averse to a little statistics in medicine provided it was realized that 'the fallacies which are incident to statistical methods in medicine impose very narrow limitations upon the useful exploitation of these methods'. Referring to the actual statistical technique employed by Pearson he remarked, sarcastically, that 'I have no doubt that the mathematical principles, in accordance with which he judges them (the typhoid statistics), are as unerring as they are completely beyond my intellectual ken'. And, as for Pearson's suggestion that tables of data should contain a 'mathematical expression for the effect exerted by the operation of chance', Wright remarked that 'every common-sense man is, even without the aid of a mathematical expression, capable of forming a judgement as to whether or not a particular result can be the result of the operation of chance'.

Wright considered that he understood very well what was required to test the efficiency of a prophylactic or therapeutic measure; the treated group of patients, a control untreated group 'which ought to correspond with the inoculated group in all points save only in the circumstance of inoculation' and a record of the exact number in each group with the percentage incidence of the disease and case mortality rate. But, in his opinion, experience taught that 'the code can never be wholly conformed to'. This pessimistic point of view seems to have been engendered by his experience with antityphoid inoculations in the army and may well, in this particular case, have been justified. He found, for example, that it was 'hardly ever practicable to obtain in an absolutely accurate manner the respective numbers of the inoculated and uninoculated' nor

'secure the exact comparability of the inoculated and uninoculated groups'. Further, it was not always possible 'to secure a
correct assignation of the sick to the inoculated and uninoculated groups respectively' and the accurate diagnosis of typhoid
fever was difficult and notoriously unreliable. Wright admitted
that with such data available the purist would never be convinced of the value of a vaccination but 'the plain, everyday
man will find it possible to reconcile the demands of his statistical conscience with the demands of practical life'. Wright's
approach was simple; where there was a patent error in any
figures, he 'allowed for it' but 'where the figures are large, all
chance errors which may have been committed in the enumer-
ation of the inoculated and uninoculated or in the classification
of the sick are spontaneously eliminated and the statistical
conclusion which may emerge may so far be accepted with
confidence.'[132]

In the minds of the medical profession at least, Wright won
the argument and antityphoid inoculation became routine not
only in the British but in foreign armies as well, with the saving
of thousands of lives. But, as Topley wrote, surveying the
immunological scene some thirty years later, 'the barriers that
separate the different departments of science are not easy to
break down, especially when the roads by which these departments are entered diverge widely in their courses and are hedged
by very different intellectual disciplines. It was natural enough
that the laboratory worker and the clinician should show little
eagerness to learn and apply the methods devised by the
statistician. But the result has been a quite unnecessary amount
of confusion; and the confusion is likely to persist so long as
the need for such methods is ignored.'[133]

Probably sometime in 1900, whilst still engaged in accumulating data on the value of prophylactic immunization against
typhoid fever, Wright's mind began to move in quite another
direction; towards the treatment of already established infections with vaccines. The reason this procedure had not already
been taken up he ascribed to the medical profession's range
of thought being limited by 'pre-suppositions'. These 'pre-
suppositions', as enumerated by Wright, appear fairly cogent
reasons why therapeutic vaccination, as opposed to prophylactic immunization was a theoretically unsound approach.

But Wright had no difficulty, by the light of theoretical reason, in disposing of them. Therapeutic vaccination ought to work, in the already infected patient, on the basis that although it might diminish the patient's resistance temporarily (negative phase) he would receive back that power 'with usury'. And had not Haffkine claimed that, sometimes, an injection of his plague vaccine aborted or reduced the severity of the disease in a patient incubating that disease? The idea seemed worth investigation but it would clearly be advisable to inform oneself of the patient's natural power of resistance and the effect produced on this by a dose of vaccine.

The first patient to be investigated was a 40-year-old man who for the previous seven years had suffered from 'furunculosis, complicated by sycosis and eczema' and had never been free from boils for more than three months consecutively. In September 1900 cultures from his boils yielded either *Staph. albus* or *Staph. aureus*. Examination of the patients blood failed to reveal any difference in its power to inhibit the growth of staphylococci as compared with normal control samples and it seemed to show even less power to agglutinate staphylococci than normal controls; to a dilution of 1 in 2 at most. Towards the end of October the patient began a course of three injections of heat-killed cultures of his own *Staph. aureus* which, although they did not enhance any growth inhibiting power of his serum, did increase its agglutinating titre to 1 in 16. However, the patient's clinical condition began immediately to improve so that, by the end of 1901, his face had been free of eruption for twelve months. At that time, the phagocytic power of the patient's leucocytes for staphylococci, *in vitro*, was measured by Major W. B. Leishman who had just developed the technique. He found that, whereas the average number of staphylococci ingested by the leucocytes of normal blood was 9.3, that of the patient's blood was 21.7 or a 'Phagocytic index' of 1:2.3. In Wright's view this lent probability to 'the assumption that the patient's continued freedom from staphylococcus invasion is the result of the inoculations undertaken'.

During the remainder of 1901 five more patients were treated. They suffered from acne, sycosis, or boils of durations from two weeks to twenty-one months. In all but one patient the phagocytic index was below unity before inoculation and in

three of them it was raised after vaccination. All improved, but not steadily, over the following weeks. But in one patient the phagocytic index was 3.3 before inoculation, after which it promptly fell to o.16 and thereafter fluctuated wildly but never regained its pre-inoculation level; but he also improved

In discussing his results Wright made no attempt to assess his evidence that his vaccinations had, in fact, contributed to the patient's recovery, nor to demonstrate a relationship between the clinical state of the patient and the ability of the leucocytes to take up staphylococci. Neither did he remark on the reproducibility, under standard conditions, of the phagocytic index which certainly fluctuated erratically from time to time in the same patient. However, he dwelt at some length on possible wider extensions of this new therapeutic approach; to conditions such as tuberculosis, erysipelas, ozoena, gleet, leucorrhoea and urinary infections.[134]

This paper by Wright has been considered in this detail because it is one of the most important publications in the history of medical bacteriology; important not as a contribution to bacteriological knowledge, for surely there are few more worthless papers in the scientific literature, but important for its influence on the development of medical bacteriology as a discipline and a profession. This paper marks the opening of an era, the era of 'vaccine-therapy' which, for good and ill, continued until the late 1940s and is not even quite passed today. The study of these six patients convinced Wright of a number of propositions (not one of which was justified on the evidence) but which rapidly became the basis of vaccine therapy; first, that vaccines produced clinical improvement, second, that the phagocytic index (soon to be refined into the opsonic index) was an accurate measure of the patient's resistance to his infection and third, that the administration of vaccines must be guided by constant reference to the phagocytic index, in particular so as to avoid putting the patient into a 'negative phase' (as had apparently happened with one of his six patients who none the less got better just as well as any of the others).

Vaccine-therapy was important in the development of medical bacteriology, because, thanks largely to Wright's forceful advocacy, it became the most fashionable method of treating

almost any disease. It is doubtful if this form of treatment produced any good results and certainly in most instances, it was valueless to the point of fraudulence. But it enormously expanded the demand for bacteriological services; to isolate the offending organism, prepare a vaccine and perform numerous measurements of the 'opsonic index' to guide the dose and frequency of its administration. Wright himself, when he moved to St Mary's Hospital in 1903, built up a flourishing department called, frankly enough, 'the inoculation department', which provided facilities for bacteriological research, handsomely rewarded himself and supported a host of research workers all on the basis of fees derived from vaccine-therapy. No hospital could afford to be without a bacteriology department and a bacteriologist; careers in the subject became possible where before openings had been very few and the status of medical laboratory workers *vis-á-vis* their clinical colleagues, was gradually raised almost to parity. Wright himself insisted that the bacteriologist should be treated as a consultant colleague and entitled to the dignities and rewards of the position. Gradually this came about and the laboratory worker won the right to 'his half of the credit and of whatever else there may be to divide'. Vaccine-therapy provided a living for many a research bacteriologist and was thus responsible, indirectly, for many of the major advances in medical bacteriology during the first half of this century; Fleming's discovery of penicillin is but the most obvious example.

In Wright's hands the scope of therapeutic vaccination was rapidly extended and good results were claimed in such conditions as chronic cystitis, appendicitis, colitis and pyelitis as well as infections of the middle ear, uterus and meninges. Particular stress was laid on the value of vaccine treatment in all local forms of tuberculosis – lupus, tuberculous adenitis, renal tuberculosis, bone tuberculosis and tuberculous peritonitis. In the treatment of tuberculosis Wright used Koch's new tuberculin, which consisted of an extract of whole tubercle bacilli, rather than the filtrate of a broth culture of the organism. Great stress was laid upon the importance of measuring the opsonic index as a guide to dosage. Wright by no means despaired of vaccine therapy in pulmonary tuberculosis and thought that it should be started when bed-rest had brought the patient's temperature

back to normal. In 1903 he first enunciated his oft-to-be-repeated slogan 'The physician of the future will, I forsee, take upon himself the role of an immunizator'. From this view Wright never departed and he used the slogan on the title page of the reprint of his book *Studies on Immunization* published in 1943 when it must have been evident that the work was of historical interest only.

Few therapeutic methods have enjoyed a vogue comparable to vaccine therapy, beginning in 1900 and continuing until well into the 1940s. Although vaccine therapy has been now quite superseded it is of historical interest to look in a little detail at this most widely practised method of treating bacterial infections, to assess its value in the relief of suffering and see in what ways it contributed to the increase of bacteriological knowledge. Its important effect on the development of bacteriology as a profession has already been alluded to. The most striking feature of vaccine therapy is the poor quality and quantity of the evidence adduced in favour of this procedure; Wright's original paper has already been examined in detail from this point of view. It must be remembered that, unlike serum therapy, vaccine therapy had no experimental basis. For example, a rabbit could be immunized with a highly virulent streptococcus and it could be shown, experimentally, that not only would this particular rabbit withstand a dose of living streptococci, which would infallibly kill an unimmunized animal, but that the serum of the immune animal would passively protect a normal rabbit against challenge with the streptococcus. In the experimental situation antibacterial serum therapy worked; the problems lay in transferring the principle to natural disease in man. These problems proved largely insoluble but a great deal was learned in their investigation. However, with regard to vaccine therapy the situation was quite different – no infected animal was ever convincingly cured of any bacterial disease by the administration of a vaccine. The evidence that vaccine therapy was a useful therapeutic procedure was derived from clinical work in man and would have been none the worse for that had it been evidence sufficient to convince a reasonable, critical and impartial observer. The trouble was that it was not. An example of the sort of evidence upon which vaccine therapy in a particular infection was recommended is worth examining. W. C.

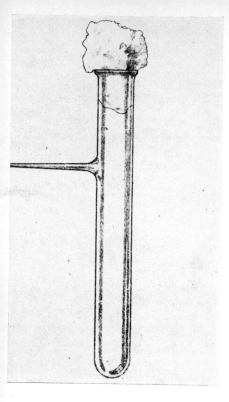

8. A tube for liquid medium which was inoculated through the side-arm which was then sealed by heat.

Woodhead and Hare 'Pathological Mycology' 1885

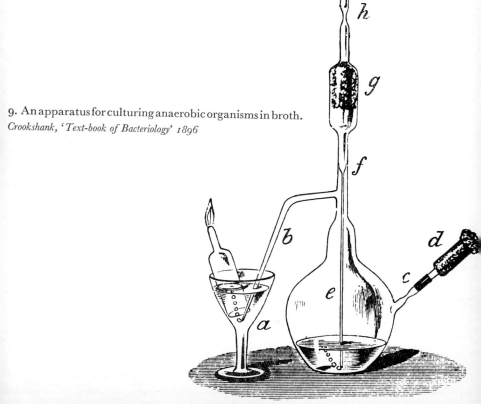

9. An apparatus for culturing anaerobic organisms in broth.

Crookshank, 'Text-book of Bacteriology' 1896

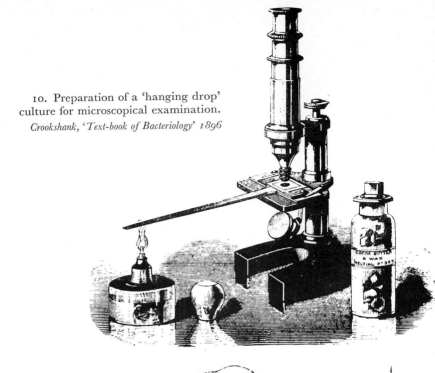

10. Preparation of a 'hanging drop' culture for microscopical examination.

Crookshank, 'Text-book of Bacteriology' 1896

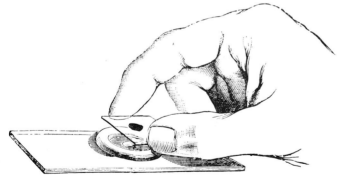

11. Examination of a 'hanging drop' culture on a simple warm stage under the microscope.

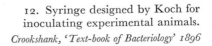

12. Syringe designed by Koch for inoculating experimental animals.

Crookshank, 'Text-book of Bacteriology' 1896

Bosanquet, a physician to the Charing Cross Hospital and J. W. H. Eyre, bacteriologist to Guy's Hospital, wrote one of the best books of the first two decades of this century on immunotherapy which went through three editions between 1904 and 1916. In the second edition, published in 1910, the evidence upon which treatment is suggested for pneumococcal infections consists of reports from the literature in which vaccine therapy had been tried, 'with benefit', in a single case of otitis media, a single case of empyema, a single case of pyaemia and two cases, by different authors, of pneumococcal peritonitis. Bosanquet and Eyre were in the habit of using vaccine therapy in lobar pneumonia and bronchopneumonia, with satisfaction to themselves, but gave no details except their favourable clinical impressions. Six years later exactly the same evidence is repeated in the third edition of their book, plus a couple of paragraphs more of similar quality. The evidence adduced in favour of vaccine therapy in pneumococcal infections is typical of the evidence upon which the whole great fabric of vaccine therapy was based, and it is not necessary to deal with all the diseases which were treated in this way.

We may conveniently jump to the year 1927, when vaccine therapy had had over a quarter of a century of extensive trial, and examine briefly an authoritative review of the subject by L. S. Dudgeon, professor of pathology, at St Thomas's Hospital. Dudgeon published a monograph entitled 'Bacterial vaccines and their position in therapeutics', based on twenty years' personal observations on the subject. He felt that he was not 'in agreement with the much too optimistic statements so frequently made on this subject' but none the less he was sure that 'No one who has studied vaccine treatment seriously can question its value'. What was the place of vaccines in therapy, according to Dudgeon? In acute generalized bacterial infections such as pneumonia, typhoid fever and malignant endocarditis, although he had occasionally seen remarkable improvement following a dose of vaccine, he, on the whole, considered the treatment useless and was opposed to its use. In infections of the urinary tract with Escherichia coli he thought that 'complete cure, following vaccine administration is uncommon' but that best results were obtained if the vaccines were administered 'after the acute symptoms have subsided'. In tuberculosis of the kidney

he regarded tuberculin as unsatisfactory and recommended surgical treatment whenever possible. In acute bacillary dysentery again Dudgeon found vaccines useless but recommended them for chronic cases. Vaccines were recommended for infections of the nasal sinuses and for recurring colds vaccines gave considerable benefit. They were, however, useless in all oral infections such as pyorrhoea and gingivitis. 'Some of the best results in vaccine therapy' were obtained in bronchitis and bronchial asthma although in bronchiectasis vaccines were useless. But 'of all infections, probably "boils" yield better results with vaccine treatment than any other and . . . the result obtained with autogenous and stock vaccines are often remarkable'. But, with regard to Wright's corner-stone of vaccine administration, the opsonic index, Dudgeon concluded, as a result of a large amount of personal work, that it was quite useless in regulating the dose nor, in fact, was there any satisfactory way of measuring the state of a patient's resistance available. In this view there was nothing novel; doubts about the accuracy, reproducibility and interpretations of the 'opsonic index' had been raised from the very beginning and it had long since been given up, if for no other reason than that it was too much trouble.

W. W. C. Topley in his book *An Outline of Immunity*, published in 1933, wrote that he was quite unable to say whether or not vaccine therapy was of any value. Such evidence as there was did not lend itself to statistical analysis and one could only await 'the final crystallization of opinion among competent clinicians'. He made the point, which is probably valid, that in some prolonged and relapsing infections, such as recurrent boils, vaccine therapy shaded into prophylaxis – a very different matter, with a sound experimental basis. Topley's treatment of vaccine therapy occupies less than two pages in a book of over 400 pages and ends with a plea for a more scientific approach to the subject: 'It does not seem unreasonable to appeal for a little more of the scientific spirit in approaching such problems as these. At the moment it seems – to some of us at least – that the practice of vaccine therapy rests on a very inadequate basis of experimental evidence. If it can be justified on the basis of clinical experience it will obviously take its place with other empirical remedies.' Thus the position of vaccine therapy, after

over thirty years of trial, was that a critical observer could not convince himself that there was any satisfactory evidence that it was of value in the treatment of disease, and yet the practice flourished. Commercial firms and public institutes of high repute published elaborate catalogues of vaccines for all occasions. Why should this have been? A number of factors tending to encourage vaccine therapy can be indicated yet they seem inadequate to account for the long life of what was almost certainly a completely useless mode of therapy. Firstly, there is the irresistible urge, in the face of a sick patient, to do something in the way of active therapy and vaccine therapy represented an advance, in a more sophisticated age, over the old-fashioned bottle of coloured water; its pseudo-scientific basis and its subcutaneous injection had an appeal to patient and doctor alike. Secondly, vaccines were easy to prepare and administer and were tolerably innocuous. Also, it must be confessed, that very rapidly a considerable commercial interest, both of manufacturers and doctors, developed. Thirdly, the types of disease for which vaccine therapy had become firmly entrenched were just the types in which the course was chronic, relapses were erratic and spontaneous improvement not uncommon. As we have seen, vaccine therapy for acute bacterial infections such as lobar pneumonia, typhoid fever or malignant endocarditis was relatively soon abandoned – the results of vaccine therapy, in these particular instances, were only too easy to assess. But, with the best will in the world, it is difficult to organize a trial of the efficacy of any remedy for recurrent boils. Lastly the blurring of the distinction between vaccine therapy for a present attack of boils or bronchitis and prophylactic immunization against subsequent attacks may have given some credit to vaccine therapy which it did not deserve.

History is, by and large, the story of mankind's progress upwards to greater and greater perfection in all things. But the progress is not, except in the long view, steady, and it is an overall progress. One civilization may be declining as another rises to even greater height, a particular period may show steady upward progress and others a flattening off or a decline. So it is with the history of bacteriology. The period 1900 to 1940 was one of overall great advance in knowledge of the subject but, to this advance, vaccine therapy contributed

nothing except to gather together wealth, a small part of which was reinvested in more profitable enterprises.

The specific treatment of diphtheria with antitoxin has already been described. But one of the greatest triumphs of bacteriology, as applied to public health, has been the virtual eradication of the disease from the enlightened parts of the world. The work leading up to this achievement will now be described. Soon after the discovery of antitoxin and its use in the treatment of diphtheria it was also used, most effectively, in the prevention of the disease among the contacts of cases. Thus, in New York, between 1895 and 1900 over 6,000 diphtheria contacts were given antitoxin of whom less than 1 per cent contracted the disease (and these mostly in the first twenty-four hours) and none of whom died. An important limitation of this valuable preventive measure was the liability of antitoxin recipients to various manifestations of hypersensitivity. The proportion of persons affected might be small, and the reactions for the most part trivial, but there were sufficient tragic deaths of healthy children from anaphylaxis to hamper its use as a public health measure. No man contributed more to the eradication of diphtheria as a public health problem than W. H. Park, of New York City.[135] Park began his work on diphtheria, as a four-year qualified doctor trying to build up a practice in rhino-laryngology, by accepting a part-time appointment in Dr Prudden's department at the College of Physicians and Surgeons of New York to study the relationship of the Klebs-Loeffler bacillus to diphtheria. This was in the year 1890 and, at that time, Prudden was sceptical about the role of bacteria in diphtheria. It was Park's duty to investigate the matter. He spent two years in the bacteriological study of severe sore throats and was able to show that, although membraneous sore throats similar to those found in diphtheria might sometimes be caused by other organisms, in true diphtheria the Klebs-Loeffler bacillus was always found. This experience in Prudden's laboratory led to Park's appointment as 'Inspector and bacteriological diagnostician of diphtheria' to the Board of Health of New York City. The diagnostic work, which was Park's main duty, led to a much fuller understanding of the bacteriology of diphtheria from a public health point of view – the accurate diagnosis in individual cases, the frequency of the diphtheria bacillus in

healthy contacts and the period during which a convalescent case might continue to carry virulent bacilli were some of the problems investigated. With such knowledge much could be done to control diphtheria in the community but, as Park himself admitted 'the total extermination of the disease under existing conditions of life here does not seem probable unless we can acquire new means to combat the disease'. Park became a vigorous advocate of the use of antitoxin prophylactically, despite the risk of reactions, but he recognized the practical difficulty involved and immediately appreciated the importance of B. Schick's discovery of a simple test to distinguish the susceptible from the naturally immune contacts of a case of diphtheria. Schick showed that, in some people, a small intradermal injection of diphtheria toxin produced an inflammatory reaction but, in others, it did not. The negative reaction was always associated with demonstrable antibodies to the diphtheria toxin in that person's blood. Testing a large number of children Schick found that whereas among new-born babies 93 per cent gave a negative reaction this proportion fell to about 50 per cent in childhood. Hardly more than six months after the publication of Schick's paper, Park and his colleagues fully demonstrated the validity of the 'Schick test' as an indicator of immunity in diphtheria. They tested 700 children admitted to the scarlet fever pavilion of the Willard Parker Hospital. Four hundred were Schick negative and, that they were immune, was demonstrated by the fact that none developed diphtheria during the period of observation. But of the 300 Schick positive children, 42 developed diphtheria. Further work showed that the use of the Schick test eliminated the necessity for administering antitoxin to two-thirds of diphtheria contacts. The age of maximum susceptibility to diphtheria, as judged by the Schick test, was found to be between the ages of 1 and 4 years and only 10 per cent of adults were susceptible. The considerable risk of diphtheria to which children admitted to hospital were exposed was illustrated by the fact that a quarter of the Schick negative children became carriers of the diphtheria bacillus.[136]

In 1913 Park resolved to try 'new means' of reducing the considerable risk of a child acquiring diphtheria when admitted to hospital. He tried to bring about active immunity using a

mixture of diphtheria toxin and antitoxin, since even small doses of the former were too toxic for practical use. That such mixtures could stimulate the production of antibody had been known since 1895 and Park himself had shown, in 1903, that this was true in horses. It was von Behring who first applied this approach to the immunization of children, in 1912, and field trials in the face of a diphtheria outbreak in Germany had yielded inconclusive results. Park vaccinated children with two or three doses, over a three- to seven-day period, of a toxin-antitoxin mixture which was still slightly toxic for guinea-pigs and assayed the patients' serum before immunization and three weeks later, for antitoxin, using the guinea-pig intradermal method. His initial results were rather disappointing. Children whose serum already contained antibody responded by producing even more, but such children were immune to diphtheria in any case; of those children whose serum contained no antitoxin, only a quarter produced significant antibody following immunization.[137] Park, however, persisted in his efforts and was eventually able to obtain a 70 per cent Schick conversion rate following two spaced doses of toxin-antitoxin mixture. By immunizing a large number of children in institutions, and therefore available for continuous observation, he showed that this was a very effective and long-lasting preventive measure against diphtheria.[138] Under Park, active immunization in New York City was pursued on a mass scale in the early 1920s and a significant fall in the incidence of diphtheria occurred at the same time. Park's campaign against diphtheria was watched with interest from other parts of the world but nowhere was Schick testing and active immunization pursued with such vigour. In Great Britain it was not until 1920 that the first paper confirming the value of the Schick test was published – the medical profession had been too overworked during the war to bother with such things. A number of factors militated against widespread active immunization with toxin-antitoxin mixtures which it required the enthusiasm of a Park to surmount. Firstly it was considered, there were other, and probably more effective, means of preventing diphtheria in the community – rapid bacteriological diagnoses, the tracing of carriers and the isolation of infected persons. Secondly the administrative and technical problems of Schick testing and subsequently

immunizing large numbers of children were formidable and parental prejudice was difficult to overcome. (One of the few attempts at mass immunization in Great Britain was made in Edinburgh, in 1924, but only 42 per cent of parents gave their consent.)[139] Lastly, there were the toxic reactions associated with immunization. The minor reaction rate was not unacceptable, but occasional highly publicized disasters were perhaps more important. Thus, in 1924, seven children died in Vienna following prophylactic toxin-antitoxin mixture; it was thought that the antitoxin had deteriorated during storage and the children had been killed by the free toxin. About the same time, in Massachusetts forty-two children had been made severely ill by vaccine which had been accidentally frozen and thawed.[140] A further objection to the toxin-antitoxin mixtures was the fact that its horse serum content might sensitize the recipient to that foreign protein.

Although good results could be obtained by means of immunization with toxin-antitoxin mixtures it is doubtful if diphtheria would ever have been completely eradicated if it had not been for the introduction of another safer and more effective immunizing agent known as 'diphtheria toxoid', in 1923. For the introduction of this immunizing agent we are largely, although not entirely, indebted to Gaston Ramon. Although he did not introduce the name 'toxoid', and indeed, throughout his life, preferred to use the word 'anatoxone'. Gaston Ramon was born in 1886, the son of a baker of Spanish descent. After a very successful school career he entered the famous veterinary school at Alfort to train for that profession. From the first his inclinations lay in the direction of the laboratory rather than clinical work and, soon after he qualified, in 1910, he was recommended to E. Roux, the director of the Pasteur Institute in Paris. Roux offered him the post of assistant in the serum production department – a job involving routine, practical work rather than research. His first research problem, suggested to him in 1915 by Roux, was to look for an antiseptic suitable for adding to antisera as a preservative. After some experiments Ramon remembered, from his days at Alfort, that formalin could be added to milk without denaturing its proteins and soon showed that this substance was a satisfactory preservative for antisera. In 1920, as a reward for his faithful service in

the serum production department, Roux put a small laboratory at Ramon's disposal for research purposes and it was here, during the next few years, that he did the work which gave him an international reputation. In 1922 Ramon described the flocculation which takes place when diphtheria toxin and anti-toxin are mixed in certain proportions and developed a simple accurate method for the assay of diphtheria antitoxin, which had hitherto always been done in live guinea-pigs. He also applied his method to the assay of tetanus antitoxin. In 1923 Ramon showed that if diphtheria toxin was rendered non-toxic, by gentle heat or treatment with formalin, it was still capable of reacting with antitoxin to give a precipitate and sug-gested that such a detoxified toxin – 'anatoxone' he called it – would be the agent of choice for active immunization against diphtheria.[141]

Ramon was not in fact the first to suggest the use of detoxified toxin as an immunizing agent. In 1915 Eisler and Löwenstein had actually used formalin-treated tetanus toxin for immuniza-tion in man, as well as trying it for diphtheria in 1921, but without obtaining a sufficiently detoxified preparation. A. T. Glenny, of the Wellcome Physiological Research Laboratories in London, in a paper published in 1921, gave a brief account of the production of satisfactory immunity to diphtheria in guinea-pigs using toxin 'changed into toxoid' by treatment with formalin rendering the product 'atoxic'. Glenny had been aware, probably for some time, that toxin which through age or contact with chemicals, such as iodine or formalin, had lost its toxicity was still antigenic and would produce antitoxin on inoculation into animals. This discovery was made accidentally because the vats, in which the diphtheria toxin was stored at the Wellcome laboratories, were washed out with formalin. It must however be pointed out that Glenny did not appreciate the possible value of 'toxoid' in active immunization in man, even though he had been actively engaged in one of the few attempts in England to immunize the children in a residential school with toxin-antitoxin mixtures. His observations on toxoid appear quite incidentally in a long review entitled 'Notes on the production of immunity to diphtheria toxin' in which he speci-fically comments that 'we do not bring forward anything that is fundamentally new'.[142] In another paper, submitted for publica-

tion in August 1923, Glenny and his colleague Barbara Hopkins stated that 'it may be possible shortly to use toxin so modified that it will be completely non-toxic without the addition of antitoxin' but a certain confusion of thought is also suggested by the statement that 'we may hope in the course of time so to improve the methods of diphtheria prevention that a single dose of modified toxin will act both as a Shick test and as an immunizing agent'.[143] Ramon, on the other hand, definitely appreciated the potential value of toxoid in the immunization of man and ended his first paper on the subject with these words:' *Cette anatoxine trouve naturellement son emploi dans l'immunization et l'hyperimmunization des animaux; de plus, grace à son inocuite et au degré tres éléve d'immunité qu'elle confers elle parait également indiquée pour le vaccination antidiphtherique de l'enfant'.*[144] In a more detailed paper, published in January 1924, he repeated these same words and added a note that a test dose of toxoid injected subcutaneously into himself had produced only a mild local reaction.[145]

Early in 1924 a quantity of toxoid, manufactured by Glenny, was sent to Park in New York for trial in man. In a small series of children it was shown that two doses of toxoid gave a 75 per cent Schick conversion rate and three doses a 94 per cent conversion rate. Park was impressed and drew attention to the advantages of toxoid over toxin-antitoxin mixtures – ease of preparation and standardization, stability and lack of toxicity. From 1925 onwards toxoid was used instead of toxin-antitoxin mixtures in his ever more energetic immunization campaign. In 1933, at an official ceremony, Park himself inoculated the millionth child to be immunized in New York City. The incidence of diphtheria continued to fall steadily until by 1940 it was no longer a significant public health problem and was indeed a rare disease.

Diphtheria immunization with toxoid was taken up on a world-wide basis but the enthusiasm with which it was applied varies very much from country to country. In some European countries immunization was made compulsory but, in France, little headway was made, except that Ramon was able to introduce the measure into the army. Really widespread immunization of children did not occur until after the Second World War. In parts of Canada campaigns organized by

J. G. Fitzgerald, D. T. Frazer and N. E. McKinnon, begun in 1927, soon produced impressive results with almost complete protection of the vaccinated and dramatic falls in the disease incidence. By the middle of the 1930s there were no cases of diphtheria in some major Canadian cities. In Great Britain a slow beginning was made and it was not until the early 1940s that mass campaigns led, gradually, to the virtual eradication of diphtheria.[146] It is interesting to note that although most people would ascribe the virtual eradication of diphtheria to mass active immunization campaigns, the direct evidence in favour of this view, particularly in the 1920s, was not wholly convincing. For example, the steady fall in the incidence of diphtheria in New York City, which occurred *pari passu* with Park's immunization activities, could have been ascribed to the mere continuation of the incidence trend which had begun long before, presumably due to a general improvement in social conditions. The Schick test provided most important evidence of the value of immunization because there was good evidence that Schick-negative persons were immune to diphtheria and active immunization definitely converted the Schick positive to negative.

The slow general acceptance of so valuable a measure as diphtheria immunization is matched by that of yet another prophylactic, now fully accepted as a very valuable agent, the establishment of which proved an even more prolonged struggle; the Bacille Calmette-Guerin (B.C.G.). It was the misfortune of both active immunization against diphtheria and tuberculosis that they were first put forward at a time when the leaders of scientific medicine were, at long last, examining more critically the evidence in favour of vaccination against infectious diseases. Looking back it seems that the pendulum swung too far in the direction of scepticism, from the earlier mood of credulity, and that the exploitation of these two measures, which have undoubtedly saved many thousands of lives, was unduly delayed. The introduction, though by no means the establishment of the value of B.C.G., we owe entirely to Albert Calmette (1863–1933) and his colleague at the Pasteur Institute in Lille, Camille Guerin (1872–1961). Albert Calmette was born in Nice, the son of a lawyer in the prefecture. His original intention was to become a naval officer but ill-health, as a young cadet, forced

him to abandon this. However, being of an adventurous disposition, he resolved to become a naval doctor. After two years training, he was sent on active service to China in 1883. In 1885 he returned to France to complete his medical education, following which he served on the west coast of Africa and in the West Indies. Calmette early showed an inclination for study and research in tropical medicine and in geographical pathology and was gradually drawn to the field of medical microbiology. In order to have better opportunities for research he transferred, in 1890, to the colonial medical service and immediately obtained leave to study at the Pasteur Institute in Paris. Here he made so favourable impression that, when Pasteur was asked by the government to suggest a suitable person to go to Saigon to set up a laboratory for the production of smallpox vaccine and rabies vaccine, he nominated Calmette. He spent two very profitable years in Saigon and, in addition to his work on smallpox and rabies, made pioneer studies on snake venoms and antisera against them which he subsequently published as a monograph. But he contracted dysentery and was compelled to return to France. Calmette was now anxious to make medical microbiology his career but the opportunities for a paid post were rare. However, through the intervention of Emile Roux, he was given the post of 'Secretaire au Conseil Superieur de sante des Colonies', a job which left him his mornings and evenings free for work at the Pasteur Institute. A few months later, the town of Lille having raised the necessary funds to set up a local Pasteur Institute, Calmette was appointed to the directorship by Pasteur. He took up his duties early on 1895 and at Lille he remained for twenty-four years and it was there that his most important work on tuberculosis was done.[147] Camille Guerin was born at Poitiers in 1872. He was trained as a veterinarian at the Alfort school and, in 1897, was recommended by Professor Nocard for an appointment on the staff of the Pasteur Institute at Lille. At first he was employed on the production of antisera but gradually became involved in Calmette's experimental work on tuberculosis. Guerin later said that, soon after his arrival in Lille, Calmette asked for his solemn promise to devote all his energies to the struggle against tuberculosis. Guerin's whole career was passed at the Pasteur Institute and for the last eighteen years of his

retirement he lived within the walls of the Pasteur Institute in Paris.[148]

Calmette was a great student of tuberculosis and, in 1920, published an important monograph 'L'infection bacillaire et la tuberculose chez l'homme at chez les animaux', which had been largely written whilst interned by the Germans during the First World War. It was his appreciation of two facts which led Calmette along the true path to an effective prophylactic against tuberculosis: first, the association between tuberculin hypersensitivity and a degree of immunity and second, that hypersensitivity only resulted from real infection with a living organism. In experiments made on cattle, with Guerin in 1907 and 1908, he showed that whereas a tuberculin negative cow always contracted fatal, acute miliary tuberculosis if inoculated intravenously with five milligrams of bovine tubercle bacilli, a tuberculin positive cow survived after exhibiting but minimal symptoms. Calmette also drew attention to an old clinical observation, Marfan's law (1886), which stated that it was very rare to find progressive pulmonary tuberculosis in individuals who had suffered from tuberculous cervical adenitis as children and recovered. He therefore sought to prepare a vaccine by attenuating the virulence of living tubercle bacilli.

Calmette and Guerin were not, of course, the first to attempt to vaccinate against tuberculosis. Many lines of work had been explored by various people since the early 1890s; perhaps the most bizarre approach being to maintain tubercle bacilli in the gut of leeches for many months, in the hope that they might become attenuated. Extracts of tubercle bacilli, killed tubercle bacilli, tubercle bacilli grown in the presence of formalin or lactic acid were all tried and found ineffective. Some degree of success was obtained by administering species of living tubercle bacilli normally foreign to the host species, such as the avian bacillus to mammals. Von Behring, in 1902, administered a strain of human tubercle bacillus, of low virulence for guinea-pigs, to cattle and showed that they had acquired a degree of immunity to challenge with the bovine strain of the tubercle bacillus. Unfortunately the cows excreted the human tubercle bacilli in their milk.

In 1908 Calmette and Guerin successfully attenuated a strain of bovine tubercle bacillus by repeatedly subculturing it,

seventy times, on a medium consisting of potato, glycerine and bile salts. They found that a calf could tolerate an intravenous injection of 100 mg. of this culture, whereas a dose of 3 mg. of the same strain, maintained on a potato-glycerine medium, without bile salts, for the same length of time, was fatal. The final assessment of the vaccine strain for cattle was necessarily prolonged since the animals had to be kept several years to make quite sure that they did not develop tuberculosis. But in 1913, trials of the effectiveness of the vaccine in cattle were being made by placing vaccinated animals and unvaccinated controls in close contact with tuberculous cattle which were freely excreting the organisms. This work was interrupted by the war, but was resumed afterwards in Paris and steps towards the vaccine in children taken.

The main preoccupation, naturally enough, was for the safety of this living vaccine but extensive work in experimental animals, including monkeys, indicated that it was harmless. Between 1921 and 1924, therefore, a careful trial of the safety of the vaccine in new-born infants was made. The vaccine was administered either by subcutaneous or intradermal injection but, because of minor local reactions which nevertheless tended to be unacceptable to parents, Calmette preferred the oral route, having shown that this was effective in experimental animals. Convinced of the safety of the vaccine, in 1924, the Pasteur Institute began mass production and, at the same time, strains of B.C.G. were made available to other countries. Soon thousands of babies, in many parts of the world, had received B.C.G.; the safety of the vaccine was amply confirmed and the problem was now to assess its effectiveness in the prevention of tuberculosis.[149]

It was at this stage that Calmette and Guerin ceased to make any effective contribution to the prevention of tuberculosis. They were content to accumulate masses of statistics, in a most uncritical manner and, in particular, regarded it as satisfactory to compare the mortality *from all causes* of vaccinated and unvaccinated children. When using the published results of others they made most careless errors, such as adding up figures in a table incorrectly, and their general ideas on conducting a controlled trial were deficient even for the period of bacteriological history concerned. Despite a general consistency of

results suggesting that B.C.G. was protective, the obvious criticisms that could be levelled at Calmette and Guerin created an atmosphere of scepticism and they themselves are, at least partly, responsible for the slow progress made with B.C.G. as an accepted prophylactic.[150]

A most important general objection was the hazardous nature of introducing a live organism into the human body and there was doubt as to whether the strain would always remain avirulent. Recent work on bacterial variation suggested that virulence and avirulence were often discontinuous phases in the reproduction of bacteria and reversion to virulence, therefore, a real possibility. Some workers even claimed to have isolated bacilli of two different colonial types from a culture of B.C.G., one of which showed some virulence for guinea-pigs. At just the time objections to the safety of B.C.G. were being made along these lines a most unfortunate accident occurred, which appeared to give grounds for the belief that reversion to virulence did occur.[151] In the spring of 1930 251 new-born babies in Lübeck were given B.C.G. by mouth and of these 72 died of tuberculosis within the first year and, in addition, 135 suffered from clinical tuberculosis but eventually recovered. There were no cases of tuberculosis among unvaccinated babies born at the same time. In fact the official investigation which followed concluded that there was no evidence that the B.C.G. strain had reverted to virulence and that the disaster had resulted from the careless contamination of the vaccine with virulent human tubercle bacilli. Two of the doctors involved were sent to prison. But the inquiry and the trial were not over until 1932 and, meanwhile, the Lübeck disaster had done great harm to the cause of B.C.G. vaccination. Calmette died in 1933 and thus did not live to see the general acceptance of his vaccine.[152]

The Lübeck disaster cast but temporary doubt on the safety of B.C.G. which was soon allayed by the testimony of the many thousands of doses administered without any untoward effects. But the proof of the value of the vaccine in the prevention of tuberculosis still lay many years ahead. Various trials of improved quality, begun in the 1930s, gradually produced a strong case in favour of the value of B.C.G. but it was not until the results of the Medical Research Council trial in Great Britain, published in 1959, became available that the value of

B.C.G. in the prevention of tuberculosis was demonstrated beyond doubt. It is satisfying to know that before this final proof became known some millions of children had received B.C.G., with the undoubted saving of many lives, and that Guerin, who died in 1961, lived to see the final acceptance of the vaccine he had first developed with Calmette half a century before.

During the first decade of the twentieth century an important immunological phenomenon of 'quasi-paradoxical character' began to be appreciated, largely due to the work of C. Richet on the experimental side and F. von Pirquet and his colleague B. Schick from the clinical point of view. The phenomenon alluded to may be conveniently named 'hypersensitivity', the complexities of which have, even today, been only partly elucidated. Manifestations of hypersensitivity of the anaphylactic type had been observed for many years but it was the work of Richet, with his colleague Portier, which drew attention to the phenomenon and they coined the name 'anaphylaxis' – meaning the 'opposite of protection', in 1902. As long ago as 1839 F. Magendie had described sudden death in dogs which had been repeatedly injected with egg albumen and, around the middle of the nineteenth century, when the infusion of animal blood into man had a brief vogue, reactions, some of which were anaphylactic in type, were noted. In 1894 S. Flexner gave a clear statement of the fundamentals of anaphylaxis in describing experiments in which animals which had been inoculated with dog's serum would succumb to a second dose, given after the lapse of some weeks, even though the dose was not lethal to a control animal. Similarly, von Behring and his colleagues noted, about the same time, that some guinea-pigs, which had survived sub-lethal doses of diphtheria toxin, might become particularly sensitive to a further injection of toxin, rather than being protected against it, and die following doses of toxin which were only a small fraction of the dose lethal to a normal animal.

With the introduction of the serum treatment for diphtheria adverse reactions consisting of fever, rashes and joint pains began to be reported and, indeed, it was soon appreciated that such reactions occurred in about 10 per cent of patients treated with serum.

The connection between these various phenomena and their general interest was not appreciated until Richet took up the subject. Richet's first acquaintance with the phenomenon of anaphylaxis occurred, in 1898, whilst studying the toxic effects of eel serum in dogs. He noted that second and third injections of eel serum caused illness in the dogs, but did not attempt to analyse the phenomenon. His real interest in this form of hypersensitivity came about, as he described, in the following way: 'During a cruise on Prince Alfred of Monacco's yacht the Prince and G. Rickard suggested to P. Portier and myself a study of the toxic properties of the Physalia found in the South Seas. On board the Prince's yacht experiments were carried out, proving that an aqueous glycerine extract of the filaments of Physalia is extremely toxic to ducks and rabbits. On returning to France I could not obtain any Physalia, and decided to study comparatively the tentacles of Actinaria which resemble Physalia in certain respects, and are easily procurable . . . while endeavouring to determine the toxic dose we soon discovered that some days must elapse before fixing it; for several dogs did not die until the fourth or fifth day after administration, or even later. We kept those which had been given a dose insufficient to kill, in order to carry out a second investigation upon them when they had completely recovered. At this point an unforeseen event occurred. The dogs which had recovered were intensely sensitive and died a few minutes after the administration of small doses.' Richet and Portier described their observations in a paper entitled 'The Anaphylactic Action of Certain Poisons', in 1902. In the following year M. Arthus described the local reaction which follows the fourth, or subsequent, injection of horse serum subcutaneously into rabbits which became known as the 'Arthus phenomenon'. He also showed that the same local reaction took place if the first injections of serum were made intraperitoneally and the subsequent one subcutaneously and, further, that death might follow the intravenous injection of serum into a rabbit which had previously received subcutaneous injections.[153]

About this time the interest of von Pirquet in the toxic effects of serum injections in children was aroused. He was working in Escherich's paediatric department in Vienna and was in charge of the scarlet fever wards. Escherich's first assistant, Moser, had

recently introduced an antistreptococcus serum for the treatment of scarlet fever. It was necessary to inject large doses, of the order of 200 c.c., so that Pirquet had abundant opportunity to study the toxic effects of which he gave a detailed clinical account, describing the onset of fever, skin rashes, lymph node swelling, etc., coming on seven to fourteen days after the injection of serum. He was, however, struck by certain odd cases in which the incubation period of the 'serum sickness' was dramatically shortened. For example, in a child who was given a second dose and developed serum sickness within twenty-four hours. In yet another case of scarlet fever, who developed serum sickness eight days after an injection of Moser's serum, and who, fifty days after the first injection, was given 2 c.c. of antidiphtheria serum prophylactically, as a case of diphtheria had been admitted to the ward, and developed immediate oedema and urticarial rash. Looking through the older hospital records von Pirquet discovered other similar cases the significance of which had not been appreciated at the time. When another case of diphtheria was suspected in the ward von Pirquet took advantage of the opportunity to compare the effect of a prophylactic dose of diphtheria serum on children who had or had not received previous serum therapy; those who had received previous serum injections reacted immediately, all but one, who had only received his first injection four days previously. Von Pirquet adduced evidence that serum sickness, and its accelerated form, were due to the presence of antibodies to the foreign serum, although he did not necessarily regard those antibodies as the same as precipitins. Indeed precipitins could only exceptionally be demonstrated in the blood of serum sickness patients. He was particularly interested in the general significance of this hypersensitivity reaction for pathology and the relation between hypersensitivity and immunity in infectious disease. This led him to a detailed study of the vaccination reaction and the tuberculin reaction. He drew attention to the fact that on revaccination the skin reaction took place much more rapidly than in primary vaccination and was more limited – hypersensitivity and immunity going hand-in-hand. His work on the tuberculin reaction led to the development of a skin scratch-test for tuberculin sensitivity which provided the first practical type of tuberculin test for field epidemiological

studies. Von Pirquet's main contributions to immunology are to be found in his classic monograph, written in collaboration with Bela Schick, *Die Serumkrankheit*, published in 1906,[154] and in a short paper, published in the same year, in which he discussed the wider implications of the hypersensitive reaction and coined the term 'allergy', to denote the general concept of changed reactivity, to be found in various states but having a similar underlying immunological mechanism.[155]

Another circumstance in which anaphylactic reactions had been noted was in the assay of diphtheria antitoxin in guinea-pigs. If guinea-pigs, which had survived an inoculation of diphtheria toxin mixed with antitoxin, were subsequently injected with a further dose of antitoxin they sometimes suffered fatal reactions. Theobald Smith drew the attention of Paul Ehrlich to this phenomenon and the latter, on returning to his own laboratory, set one of his colleagues, Otto, to study the problem. About the same time two Americans, Rosenau and Anderson, of the Laboratory of Hygiene in Washington, took up the subject independently. Otto published his work, in 1906, under the title of *Das Theobald Smitchsche Phaenomen* and his work, together with that of Rosenau and Anderson established a number of new points.[156] It was shown that the actual diphtheria toxin was not an essential to the reaction, although it did seem to potentiate it, and that the reaction was specific so that a horse serum sensitized guinea-pig reacted to a dose of horse serum only. It was found that only a small dose of antigen was needed to sensitize an animal and, in fact, repeated injections of large doses prevented the development of anaphylaxis. It was found that a definite incubation period, of about ten days minimum, was necessary for sensitization to take place. As regards the mechanism of anaphylaxis, it was correctly ascribed to the development of an antibody, and the subsequent combination of that antibody with antigen which somehow gave rise to the symptoms. At first it was thought that anaphylaxis resulted from the union of antigen and antibody in the circulation but, when it was shown that a normal animal could be passively sensitized to anaphylaxis by the injection of the serum of a sensitized animal, it was soon appreciated that an interval of time must elapse between the injection of the sensitizing serum and the shocking dose of antigen. This suggested that the

antibody had to become fixed to cells. Incontrovertible proof of the cellular site of the anaphylactic reaction was brought by Schultz, in 1910, when he demonstrated the *in vitro* contractions of small intestine of thoroughly washed pieces of the small intestine of sensitized guinea-pigs when brought in contact with antigen.[157] This technique was subsequently extended and improved by H. Dale, in 1913, using strips of smooth muscle from the uterus rather than the intestine and this phenomenon is, today, known as the 'Schultz-Dale' reaction.[158]

Yet another hypersensitive state which began to be studied in the first decade of this century was hay-fever, although it was not until about 1917 that it was appreciated that the disease had an immunological basis. The connection between catarrhalis aestivus and grass pollen had been established in the nineteenth century, largely due to the classic work of C. H. Blackley, who published his *Experimental Research on the Cause and Nature of Catarrhus Aestivus* in 1873. His work formed the starting point of the immunological investigations some thirty years later. An American, W. P. Dunbar, who was educated in Germany and became director of the Hygiene Institute in Hamburg, began his studies about 1900. He thought that the grass pollen contained a toxin which was responsible for the symptoms of hay-fever and for the reaction produced by pollen extracts pricked into the skin. He also showed that pollen was antigenic in animals and that an antiserum could be manufactured against it. Dunbar made an antiserum, which was marketed under the name of 'Pollantin', for local application to the eyes and nostrils of hay-fever patients to neutralize the toxin. It appeared to have some beneficial effect. About 1908, L. Noon, working in A. E. Wright's laboratory, conceived the idea of actively immunizing patients with pollen extracts, the rationale being the same as Dunbar's passive immunization against the 'toxin'. Noon, wishing to have some objective measure of his immunization's effect, developed a standardized test of the patient's reactivity to the toxin by instilling different dilutions of an aqueous pollen extract into the eyes of hay-fever patients. Highly sensitive patients reacted to a high dilution, with a reddening and itching of the conjuctiva, i.e. with but a few Noon units of pollen; less sensitive patients required a larger dose to produce an effect and normal persons did not react at all, even

with 20,000 units ml. In 1911 Noon, who was by this time dying of tuberculosis, published a preliminary note reporting the results of immunization with pollen extracts, in three hay-fever patients. He showed that following immunization the concentration of pollen extract required to set up a conjunctival reaction was much increased.[159]

Noon's work was taken over by his colleague, J. Freeman, who, in September 1911, reported the results of a trial on twenty hay-fever patients, both with regard to their increased 'resistance' to pollen instillation in the eye and their clinical condition. Immunization always increased resistance in the conjunctival test and nearly all the patients claimed to be significantly less affected during the hay-fever season. Freeman's work hardly amounted to an exhaustive trial of the method but vaccination of all kinds were fashionable and the St Mary's Inoculation Department promptly put their pollen vaccines on the market through Messrs Parke Davis & Co.[160] As has been indicated, at this time, the mechanism of hay-fever was considered to be an immunization against a toxin. It was Wolff-Eisner who first suggested that the symptoms of hay-fever were a manifestation of anaphylaxis-like hypersensitivity.[161] This view at first did not make much ground because of the difficulty of sensitizing experimental animals to pollen extracts or detecting precipitating or complement fixing antibodies but, in the 1920s, these difficulties were overcome by a number of different workers.

This account of the classical phase in the study of immune hypersensitivity may be rounded off with a consideration of a paper, published in 1921, by C. Prausnitz and H. Kustner, of the Hygiene Institute in Breslau. They took the opportunity to study in a most beautiful manner their personal afflictions, for Kustner was highly sensitive to fish and Prausnitz suffered from hay-fever. They showed that, although they could not detect precipitating or complement-fixing antibodies in their serum to either fish or pollen, it was possible passively to transfer Kustner's sensitivity to fish to Prausnitz, by the intra-dermal injection of o.1 ml of the former's serum, but that it was not possible to transfer Prausnitz's sensitivity to pollen to Kustner in the same way, nor would Kustner's serum passively sensitize the skin of guinea-pigs.[162]

7 The Main Developments in Bacteriology during the Early Twentieth Century

By the year 1900 the microbial cause of most of the important bacterial diseases of man such as typhoid fever, cholera, plague, undulant fever, lobar pneumonia, cerebrospinal meningitis, gonorrhoea, tuberculosis, leprosy, tetanus and diphtheria were known. During the following ten years the bacterial cause of a few more important diseases, such as syphilis and whooping cough, were isolated but the major advances of the first decade of the present century lay in the accumulation of a more detailed knowledge of some of the important pathogens and in the clarification of the part played by organisms such as the streptococcus, staphylococcus and coliform organisms, which, although known by the beginning of the century, played a larger part in human pathology than had, at first, been supposed. Another field in which important advances were made was the epidemiology of bacterial disease. Advance in the treatment of bacterial disease was on the whole, disappointing and no further triumphs comparable to the antitoxin treatment of diphtheria were achieved. But it was the attempts at therapy, more than any other factor, which led to an increased understanding of the biology of such important organisms as the pneumococcus, meningococcus and the streptococcus. Another line of work which yielded fruitful results was the attempt to develop specific diagnostic methods in clinical bacteriology. Such work, for example, extending from the serological diagnosis of typhoid fever, gradually disclosed the whole range of human pathogens now known as the Salmonella. The present chapter will be devoted to some of the discoveries made in ways just indicated.

A good example of the rich harvest of knowledge gained in

this way is the growth of fundamental knowledge which stemmed from attempts to treat pneumococcal pneumonia with antisera. The first to attempt to treat pneumococcal pneumonia by an antiserum were G. and F. Klemperer, in 1891, within less than a year of Behring's discovery of antiserum treatment of diphtheria. Their approach was, however, crude, although they appeared to produce an antiserum in rabbits which had some protective effect experimentally, and they treated six human cases and were favourably impressed with the result.[163] At first their work could not be confirmed. But, in 1896, J. W. Washbourn showed that the serum of rabbits rendered immune by inoculations of heat-killed pneumococci, gave a variable degree of protection to normal rabbits if injected soon after a fatal dose of pneumococci had been administered. He also found that, in one case, the serum of a human patient convalescent from pneumonia had a partial protective effect in rabbits. Washbourn also examined the question as to whether or not varieties of the pneumococcus existed but, himself, had 'never succeeded in obtaining distinct varieties existing as constant types'.[164] Washbourn then immunized a pony, first with dead and then with living pneumococci, and showed that its serum developed powerful protective properties for rabbits, if administered within a few hours of the pneumococci, even though the challenge pneumococci were from a different source from the vaccine strain. He was careful to use a standard highly rabbit-virulent strain for challenge purposes. Having shown that the immune pony serum was innocuous in large doses in rabbits, he tried the therapeutic effect of his serum in two cases of pneumonia in man. In both cases the temperature fell after administration of the serum and the patients made a good recovery.[165]

Meantime a Dr Pane, in Naples, had been manufacturing an antipneumococcal serum, in a cow and a donkey, in large quantities, which he was willing to supply. Washbourn obtained some of this serum and confirmed that it protected rabbits very well and pointed out that it was of special interest that this serum would protect against pneumococci from two different sources. He felt sure that antipneumococcal serum would be a successful therapeutic agent in man.[166] It was already appreciated that antisera against streptococci would protect against some strains but not others, apparently identical, so Washbourn,

with his friend, J. W. H. Eyre, thought it important to test a single antipneumococcal serum against a number of strains of pneumococci from definitely different sources. They tested Pane's serum against five strains; distinct protective effect was evident against four of them but none against one. They therefore concluded that 'there must be varieties of the pneumococcus which in morphology, cultural characters and virulence are similar, but have other more subtle differences'.[167] It is worth noting that Washbourn was well aware that his antiserum was not bactericidal; he showed that pneumococci would grow in it, but clumped together at the bottom of the tube. Mennes showed very well the part played by antiserum in immunity to pnemococcal infection when he demonstrated that whereas leucocytes were unable to phagocytose pneumococci suspended in normal serum they did so from immune serum.[168] Denys confirmed this, in 1897, writing that, without the assistance of immune serum, the leucocytes found themselves 'disarmed'. Thus the part played by serum factors, later to be known as opsonins, was appreciated some four years before A. E. Wright's work on this subject. Antipneumococcal sera continued to be used in the treatment of lobar pneumonia and the journals contain very numerous reports, usually of single cases or very small numbers of patients, treated in this way. Many reports suggest that the antiserum was useful but, in 1904, when J. M. Anders performed the useful task of reviewing the subject, he concluded that the reduction in mortality achieved with serum therapy was so small as not to justify its use.[169]

An important student of the pneumococcus during the first decade of the twentieth century was F. Neufeld, who worked under Koch in Berlin, and eventually became director of the Institut für Infektionskrankheiten. He described, in 1902, the phenomenon of the apparent swelling of the capsule of the pneumococcus when it was mixed with specific antiserum (the 'quellung' reaction) which was later to be used as a rapid method for typing pneumococci prior to serum therapy. He also developed a mouse protection test and showed that some antisera, whilst protecting against one strain of pneumococcus, were totally ineffective against others. [170, 171]

The importance of lobar pneumonia, a common disease carrying an average case mortality of about 25 per cent, led to

the disease being taken up as a project by the Rockefeller Institute, about 1912. During the next ten years the history of pneumococcal pneumonia centres around the Rockefeller Institute and Hospital in New York and the work of five men; R. Cole, A. R. Dochez, L. J. Gillespie, O. T. Avery and M. Heidelberger. The results of their joint work may be summarized as follows: Using the mouse-protection test it was shown that the pneumococci isolated from cases of pneumonia, for the most part, fell into three well-defined sero-types and in less than a quarter of cases were the pneumococci of miscellaneous serotypes. It was further shown that the pneumococci which could commonly be isolated from the throats of healthy persons did not belong to the disease-producting serotypes, but that healthy contacts of cases of pneumonia often harboured the virulent strains in their throats. In 1917 the Rockefeller group reported their results of treatment with specific antiserum in 107 cases of Type I lobar pneumonia. Their case mortality was only 7.5 per cent but no properly controlled trial was done, the mortality of treated cases being compared with the mortality, in the same hospital, in the years before specific serum therapy, which was about 25 per cent. But the well-known, great variation in pneumonia mortality, in different places and at different times, made the Rockefeller groups results less convincing than they might have been. In the course of their studies the Rockefeller workers noted that pneumococci produced, in culture, a 'soluble specific substance', which they showed to be a carbohydrate, which was type-specific, whereas the protein fraction of the penmococcus was only species-specific. Further work led them to associate the carbohydrate specific substance with the virulence of the organism thus throwing new light on the mechanism of pathogenicity of the pneumococcus. Subsequent work has shown that carbohydrate capsules are virulence factors in a number of other species of bacteria.[172] But, in the long term, by far the most interesting discovery emanating from this mass of work on the serology of the pneumococcus stemmed from the work of a British bacteriologist, F. Griffith. Griffith, who was killed in an air-raid, at the age of 60, in 1941, was a graduate of the University of Liverpool and the brother of a distinguished bacteriologist, A. S. Griffith. F. Griffith joined the Local Government Board as a bacteri-

ologist, in 1910, and remaining in the same post when the Board was taken over by the Ministry of Health, worked under the primitive conditions provided by the Ministry until his death. For over thirty years of his working life Griffith followed a single line; he believed that advances in epidemiology of infectious disease could only come through a more detailed knowledge of the causative organisms themselves, particularly strain differences within a single species. He worked on the differentiation of types among meningococci and staphylococci and his work on the typing of streptococci forms the basis of the modern typing system which has enabled the epidemiology of the various streptococcal diseases to be worked out.[173] Griffith's work on the pneumococcus had a similar origin – an interest in differentiating types within a species.

Griffith's paper on 'The significance of pneumococcal types', published in 1928, has been described as 'the fuse of a time bomb whose explosion sixteen years later ushered in the greatest revolution in biological knowledge of the twentieth century'.[174] But, at the time, neither Griffith nor anyone else recognized this and the paper was presented as a contribution to the epidemiology of pneumonia and, in this particular respect, it is probably not of any significance. None the less it is worth considering in detail.

As part of his routine work Griffith had to examine large numbers of sputum samples from patients suffering from lobar pneumonia, isolate the infecting pneumococci and divide these, by serological means, into Types I, II and III and the heterogeneous Group IV. He had noticed, over the years from 1920, a distinct diminution in the proportion of cases of pneumonia due to Type II pneumococci, and an increase in those due to Group IV. Also, occasionally, more than one type of pneumococcus would be isolated from the sputum of a case of pneumonia. Griffith thought that these observations might be explained in one of three ways; the patient, who previously carried a number of different strains, became infected with one of them or the patient carried one particular strain which, by mutation, developed into the disease-producing strain or, thirdly, that the second type of pneumococcus appeared, as a response on the part of the infecting organism, to the patient's resistance mechanisms. From the epidemiological point of view these

were important points and his subsequent experimental work was designed to investigate them. Was it possible that one strain of pneumococcus might, in certain circumstances, be transformed into a different serotype? Griffith decided to see whether or not a rough, avirulent pneumococcus (which he could obtain from a smooth virulent strain by growing it in specific antiserum or by repeated subculture) could be transformed into a smooth virulent organism. He injected a mixture of living, rough organisms with heat-killed, smooth organisms into mice and showed, quite conclusively, that the mice died of infection with living, smooth pneumococci. He showed further that although a rough, avirulent organism could be most easily transformed into a smooth, virulent organism of the same type from which it had been derived, it was also possible to transform one type of pneumococcus into a different type by the same method. Griffith was not particularly concerned to analyse the mechanism behind the transformation phenomenon but supposed that it was affected, not by the capsular polysaccharide itself, but by 'that specific protein structure of the virulent pneumococcus which enables it to manufacture a specific soluble carbohydrate'. He was not struck by the unusual nature of this 'specific protein structure', which withstood heating to a degree which denatures all normal proteins. He thought that the evidence he had brought forward was not inconsistent with the idea that pneumococci might contain major and minor capsular antigens and that the development of pneumonia might be due to 'the evolution in the individual of special types most suited to set up pneumonia'. Likewise, as the patient recovered, the disease-producing type might revert to the heterogeneous Group IV organisms which were found in healthy persons. There was thus a regular sequence of changes in type of pneumococcus before the development of pneumonia and during recovery from it.[175]

By the time that Griffith was killed the importance of his discovery of pneumococcal transformation for biology as a whole was not appreciated, either by himself or anyone else. In 1933 J. L. Alloway had shown that the transforming principle was present in the cell-free extracts of the donor strain and that transformation would take place *in vitro* – as well as in mice. By 1944, the systematic work of Avery and his colleagues

had demonstrated that the transforming agent was deoxyribonucleic acid and the way was opened for the understanding of the chemical basis of heredity.

There was yet another fruitful extension of the Rockefeller Institute's studies on the pneumococcus, to the work of R. Dubos and his discovery of the antibiotic tyrothricin in 1939. This discovery in many ways more truly represents the beginning of the antibiotic era than Fleming's discovery of penicillin ten years earlier. It was known that the pneumococcus capsular polysaccharide accumulated in the infected animal or was excreted, but that none appeared to be broken down by the host's enzyme systems. Dubos set about systemically looking for enzymes in other organisms which would attack pneumococcus polysaccharide and, in a bacillus isolated from soil, found one which would specifically degrade Type III polysaccharide only. This enzyme had some protective effect in both mice and rabbits against pneumococcus Type III infection. This work formed the starting point of the search, by Dubos and others, for antibacterial substances in the micro-organisms of the soil which has since provided most of the antibiotics used in clinical practice.[176]

By the early 1940s all the enormous mass of work which had been done in an attempt to develop an effective serological treatment of pneumonia was rendered obsolete by the introduction of chemotherapy with sulphonamides and penicillin. Serotherapy of pneumonia had been almost a total failure. But some lives had undoubtedly been saved and much had been learned of general importance for medical bacteriology and biological science as a whole.

In the same way as with the pneumococcus, the even less successful attempts at serum therapy for streptococcal infections led to important results. We have seen that the streptococcus was one of the earliest organisms shown to play a part in human disease and its connection with purpural fever, erysipelas and wound infections was well known by the 1890s. Likewise the streptococcus was strongly suspected of playing an aetiological role in scarlet fever. In 1895 A. Marmorek gave an account of his attempts to produce an antistreptococcal serum.[177] He immunized horses with broth cultures of streptococci and showed that the resulting antiserum had a definite protective

effect in laboratory animals. His results with various sorts of human streptococcal disease, including scarlet fever, appeared encouraging, with abatement of toxic symptoms and fall in temperature. P. Moser, unable to obtain good results with Marmorek's serum in the treatment of scarlet fever manufactured his own by immunizing horses, with several different strains of streptococcus, and was convinced of the beneficial effects on the general condition and the rash in scarlet fever, although not sure that it made any difference to the localized pyogenic manifestations.[178] He also found that antisera prepared in rabbits against scarlet fever strains of streptococcus agglutinated all scarlet fever streptococci but not those from other conditions, but other workers were unable to confirm this. Indeed attempts had been made to use an agglutination reaction for the diagnosis of streptococcal disease, in a manner analagous to its successful use in typhoid fever, as early as 1897. Positive results were obtained by a number of workers but agglutination of streptococci proved feeble and irregular. G. H. Weaver[179] attempted to reduce the matter to some order employing sixteen different strains of streptococci isolated from a variety of human infections and serum from fifty-two individuals suffering from streptococcal infections as well as from two typhoid and four normal individuals. His results only served to emphasize the difficulties of the subject; he showed that the pH of the broth, in which a strain of streptococcus was grown, might determine whether or not it was agglutinated by a single serum and that serum from patients not apparently suffering from streptococcal infections commonly agglutinated streptococci.

Meanwhile attention had been drawn to the haemolytic power of streptococci and of its possible significance in classification. In 1902 Marmorek noted an 'elegante aureole d'hemoglobine dissonte' around stretptococcal colonies on agar containing a little rabbit blood. But it was H. Schottmüller, whose attention had been drawn to the haemolytic power of streptococci by his work on blood culture, a technique he introduced in 1897, who was the first to attempt to classify streptococci by the haemolysins produced. He identified a 'Streptococcus longus pathogenis' which produced a zone of complete clearing around colonies on human blood agar and distinguished it from

a 'Streptococcus mitior seu viridans' which produced a greenish change, similar to that produced by the pneumococcus, but less intense and a 'Streptococcus mocosus' which produced only a darkening of the medium.

Many workers subsequently studied the haemolytic effect of streptococci on blood agar using a variety of techniques, different species of blood and streptococci from a wide variety of sources both human and animal. Shottmüller's observations were, in general, confirmed with numerous minor variations which led to a plethora of cumbersome names for different strains of streptococci. J. H. Brown, a pupil of Theobald Smith, took advantage of an investigation into the epidemiology of an apparently milk-borne outbreak of tonsillitis, in which numerous strains of streptococci were isolated from the throats of normal individuals, tonsillitis cases, suspected milk and cows, to accumulate a large number of strains of the organism. Over the following five years he studied their cultural characteristics and fermentation reactions and proposed, in 1919, the modern, basic classification of streptococci according to their haemolytic activity. He introduced the term 'alpha haemolytic' for streptococci producing the greenish 'viridans' change, preferring this term for simplicity and because the degree of green change was so variable. Strains causing complete haemolysis he designated 'beta haemolytic' and those which produced neither change 'gamma' strains. There was no particular reason for the choice of names except that the 'alpha' strains were the first noted and found on all throat swabs whereas the 'beta' strains were found particularly in association with disease, notably in secondary pyogenic manifestations of scarlet fever and which were considered to be aetiologically related to the epidemic.[180]

A good deal of the interest in the serology of the streptococci was removed by a paper by G. Jochmann, published in 1905, in which he brought forward cogent reasons why streptococci were not the cause of scarlet fever. He pointed out that scarlet fever, in its early stages, was a highly specific disease quite unlike the suppurative diseases commonly associated with streptococcal infection, and that streptococci could not be isolated from the blood early in the disease. Jochmann accepted that many of the complications of scarlet fever were due to streptococcal infection and was puzzled by the good effects brought about by Moser's

antistreptococcal serum. Largely through Jochmann's influence the idea that streptococci were the cause of scarlet fever was more or less abandoned for about thirteen years.[181]

Towards the end of the First World War the workers at the Rockefeller Institute, having completed their work on the serotherapy of pneumonia, took up the study of the haemolytic streptococci with the same objective and the same basic approach. Dochez and Avery were joined by a colleague whose name is for ever linked with the streptococcus, Rebecca Lancefield. Using agglutination tests, with sera prepared in rabbits, and mouse protection tests the Rockefeller workers soon showed that haemolytic streptococci would be divided into distinct serological types and the idea, originally put forward by Moser and Pirquet in 1902, that streptococci from scarlet fever cases belonged to a special strain gained ground again for a few years. George and Gladys Dick in a series of papers published between 1921 and 1924 may be said to have finally proved that the haemolytic streptococcus causes scarlet fever. They swabbed the throats of human volunteers with streptococci from a case of scarlet fever and reproduced the typical disease. But even here confusion was introduced because only a minority of their volunteers developed scarlet fever, although some developed tonsillitis but no rash. The Dicks themselves considered that their experiments did not justify the conclusion that all cases of scarlet fever were due to a streptococcus.[182] But Dochez had also obtained strong evidence in favour of the haemolytic streptococcus as the cause of scarlet fever. He found that this organism could be constantly isolated from the throats of patients in the early stages of the disease and that all such strains appeared to be of the same serological type. Further, he prepared an antiserum for cultures of scarlet fever strains in a horse and showed that this antiserum would cause local blanching of the rash in scarlet fever.[183] In this last procedure Dochez was taking advantage of an observation, published in 1918 by W. Schultz and W. Carlton, in which they showed that serum taken from a convalescent case of scarlet fever, but not serum taken early in the disease, would, if injected into the skin of another case of scarlet fever, cause local blanching of the rash.[184] As the association between scarlet fever and a streptococcal infection of the throat became established it became an increasingly pressing

problem to discover whether or not a special strain of streptococcus was the cause of other conditions such as puerperal
fever. A great deal of serological work was done based on the
agglutination of suspensions of streptococci in tubes. One of the
most important workers in the field of streptococcus typing was
F. Griffith who introduced two technical methods which went
some way to making typing more accurate. The difficulty with
tube agglutination tests with streptococci lay in the instability of
streptococcal suspensions, and this Griffith partly overcame by
testing for rapid agglutination by antiserum on a slide – agglutination could be seen quite easily even if the original suspension
was rather granular. He also employed an agglutination
absorption test in which unknown strains of streptococci were
tested for their ability to absorb out the agglutinins from a
serum against another particular streptococcus strain. Using
these methods Griffith showed that thirty-seven out of eighty-
one strains of streptococci from cases of scarlet fever belonged
to three serotypes, the remaining strains being serologically
heterogeneous. By contrast, streptococci from cases of puerperal
fever were mostly serologically heterogeneous, although some
strains were apparently similar to scarlet fever strains.[185] The
classification of the haemolytic streptococci was thus in the
1920s in a state of confusion and remained so until, after many
years work on the subject, the situation was clarified by the
work of Rebecca Lancefield. She abandoned the agglutination
methods and found that immune sera could be prepared which
reacted with an extract of streptococcus giving a precipitation
test. Using this method she divided streptococci into five groups
which bore a definite relation to the source of the culture; most
strains from human disease including scarlet fever belonged to
the single group A.[186] Griffith, employing his rapid slide
agglutination technique, showed, in 1934, the Lancefield group
A streptococci including those from scarlet fever, could be
divided into at least twenty-seven different serotypes.[187] The
way was now open, by combining Lancefield grouping and
Griffith typing, not only to distinguish pathogenic from non-
pathogenic strains of streptococci, but to distinguish, for
epidemiological purposes, strains among the pathogenic Group
A.

Another important disease in which the role of the haemolytic

streptococcus was not appreciated until the 1930s was rheumatic fever. An association between tonsillitis and subsequent rheumatic fever was clearly noted by J. K. Fowler in 1880. His attention had first been drawn to the association when he himself suffered from rheumatic fever in 1874. Since that time he made a point of inquiring for a history of recent tonsillitis in all his rheumatic fever patients and obtained a positive history in about 80 per cent.[188] This association was confirmed over the years and there were even epidemics in closed communities of tonsillitis followed almost immediately by epidemics of rheumatic fever. Moreover, patients who had had rheumatic fever commonly experienced a recrudescence of their rheumatism following an attack of tonsillitis. Among the first to insist on the crucial part played by the haemolytic streptococcus in rheumatic fever was A. F. Coburn.[189] Attempts to isolate streptococci from cases of rheumatic fever were generally unsuccessful, but not completely so, and a number of workers reported the isolation of streptococci from the blood and joints. But it was not until the exploration of the indirect approach – looking for antibody to the streptococcus and its products in the serum of cases of rheumatic fever – that the relationship became settled. An early report of antistreptococcal antibodies in the blood of rheumatic fever patients was made by B. Schlesinger and A. G. Signy, from the Hospital for Sick Children in Great Ormond Street.[190] They tested the serum of twenty-one cases for precipitins against a saline extract of ground-up streptococci, both haemolytic and viridans types. In fourteen cases, from whom a haemolytic streptococcus had been isolated, precipitins were present in six. Moreover, precipitins either to extract of haemolytic or viridans streptococci were present in four out of seven. Independently and almost simultaneously Coburn, in New York, obtained similar results. The full significance of the haemolytic streptococcus was, even so, not appreciated and Schlesinger and Signy thought that infection with streptococcus viridans as well as the haemolytic streptococcus might be important. It was an improvement in the technique for detecting antistreptococcal antibodies, devised by E. W. Todd, which finally settled the role of the streptococcus in rheumatic fever. Todd developed a method for titrating antibody to the haemolysins of the streptococcus, having shown that these were species

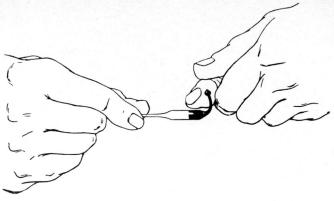

13. Collecting blood from a thumb prick into a Wright's capsule
for serological investigation.

Wright, 'Technique of the Teat and Capillary Glass Tube' 1912

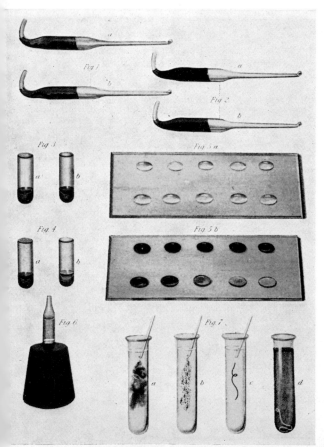

14. Serological
investigations on the small
quantities of blood
collected in a Wright's
capsule.

*Wright, 'Technique of the Teat
and Capillary Glass Tube' 1912*

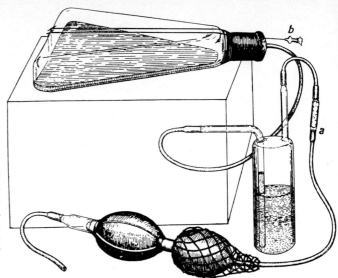

15. Apparatus for the culture of bacteria in quantity for the preparation of vaccines.

Wright, 'Technique of the Teat and Capillary Glass Tube' 1912

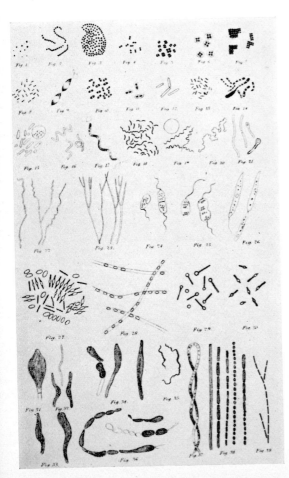

16. Microscopical appearances of different species of bacteria.

Crookshank, 'Text-book of Bacteriology' 18

specific, being produced by all strains of haemolytic streptococci, but distinct from the haemolysin of the staphylococcus and pneumococcus. He found, in a large number of patients with rheumatic fever, that those in an acute attack had markedly more antistreptolysin antibody in their serum than cases which were quiescent.[191] A paper by W. K. Myers and C. S. Keeper, published in 1934, confirmed Todd's findings and extended them in a number of ways; they showed that the antistreptolysin titres of rheumatic fever patients were of the same order as those found in patients with overt streptococcal infections and much higher than normal controls.[192]

Our knowledge of the Salmonella group of organisms grew out of experience in using immunological techniques both for the identification of bacteria isolated from patients and the complementary method, of examining the patient's serum for antibody to bacteria which were suspected of causing his illness. The story of our understanding of the Salmonella is long and complicated for, from the first isolation of a number of this group to our modern appreciation of the group and its inter-relationships, there is a period of about fifty years. The very many scattered relevant observations were only finally gathered together in the 1930s. The story is, however, an interesting one and is very characteristic of the development of bacteriological knowledge in the period immediately following the 'golden age' of the 1880s and 1890s, when the bacterial causes of so many important diseases were being rapidly discovered.

In trying to reconstruct the history of our knowledge of the Salmonella various starting points might be chosen but, from the point of view of human pathology, perhaps Gärtner's discovery of the organism now called *Salmonella enteritidis* is as convenient a point as any. In 1888 Gärtner, the 40-year-old professor of hygiene at Jena, investigated an outbreak of gastro-enteritis attacking fifty-eight persons who had eaten the meat of a cow which had been slaughtered because it appeared diseased. From the tissues of the cow and from a fatal human case he isolated what he thought was a characteristic bacillus and named it 'Bacillus enteritidis'. During the following ten years a number of other outbreaks of gastro enteritis, from which similar organisms were isolated, were described from various parts of Europe. But it was not until Grüber and

Durham described the phenomenon of specific agglutination, in 1896, that it was possible to determine the similarity or the difference of these organisms with any certainty. Durham himself was the foremost investigator of what was known, at the time, as 'meat poisoning' in England. By 1898 he had investigated four outbreaks amounting, in all, to 256 cases with nine deaths. The organisms which he isolated from the infected meat and tissues of fatal cases were serologically identical with Gärtner's bacillus. Moreover, a proportion of survivors showed small amounts of agglutinin for Gärtner's bacillus in their serum.[193] In 1899 there was an outbreak of meat poisoning in the Belgian village of Aertryck which was important in that the organism isolated, although similar to Gärtner's bacillus, was serologically distinct.

Meanwhile, now that serological methods had provided an exact method for identifying the typhoid bacillus, it was being shown that, not infrequently, a bacillus, very similar to the typhoid bacillus but inagglutinable by antisera against the typhoid bacillus, was isolated from the urine or blood of patients suffering from a typhoid-like illness. For these organisms, in 1900, Schottmüller coined the name 'bacillus paratyphosus'. About this time K. Landsteiner, working in Grüber's laboratory, showed that a rabbit anti-typhoid serum would also agglutinate Gärtner's bacillus and Durham confirmed this using the serum of human patients recovered from typhoid fever. However, the typhoid sera could be diluted so that they ceased to agglutinate Gärtner's bacillus whilst still agglutinating the typhoid bacillus.[194]

The general similarity of the typhoid and paratyphoid organisms, the meat-poisoning organisms and the coliform organisms like those described by Escherich was appreciated and, in 1900, Durham gave a useful review of the state of knowledge of the whole family of bacteria now known as the Enterobacteriaceae.[195] The range of characters available to Durham was limited; morphology, colonial appearances, motility, gas production, changes induced in litmus milk whey (in his opinion a most valuable differential medium) and serology gave a basis for distinguishing this group of bacteria. He divided the whole family into three main divisions: (1) Typhoid-like, (2) Colon-like, and (3) Bacillus lactis-aerogenes-like. The sub-groups within each of these main divisions brought together, for the most part,

organisms which today are regarded as clearly related. Durham had been the first to put fermentation reactions, as a means of distinguishing bacterial species, on a practical basis but he appreciated that a much wider range of substrates was desirable. With this object in view he made extracts from a wide range of vegetable material, truffles, yeast, artichoke, acorns, coconut, mistletoe, bananas, etc., and tested the power of bacteria to ferment them, but had little practical success. His work in this field was given up when he went to Brazil, in 1900, to investigate diseases of the tropics.

In 1902 a test that was to prove of the greatest importance in the development of our knowledge of the Salmonella was introduced by A. Castellani. Castellani was, at the time, a newly qualified doctor doing a few months post-graduate work in bacteriology under Professor Krause at Bonn. He was investigating the possibility of serological diagnosis of mixed infections, a not very likely occurrence, using rabbits infected experimentally. He showed that the serum of a rabbit which had been inoculated with typhoid bacilli developed agglutinins, not only for the typhoid bacillus, but also for *Escherichia coli*. However, if the rabbit's serum was mixed with a heavy suspension of typhoid bacilli it lost its power to agglutinate both organisms. On the other hand, if a rabbit was simultaneously inoculated with both *Salmonella typhi* and *Escherichia coli* it required heavy suspensions of both organisms to remove all the agglutinins.[196] This test, of little significance in the context investigated by Castellani, was used by A. E. Boycott, in 1906, as a means of distinguishing serologically organisms which were agglutinated, not only by sera prepared against them, but also by sera against other related organisms. Boycott showed, for example, that if he took the serum of a human typhoid which agglutinated also Gärtner's bacillus, the Aertryck bacillus and others, and absorbed it with typhoid bacilli it lost all its agglutinating power, but, absorption with the other organisms removed the agglutinins active against that organism but not against the typhoid bacillus. The test could, therefore, be used to determine, in an infallible way, the identity or non-identity of two organisms; if two organisms were the same the second organism should remove completely the agglutinating power of a serum against the first. These laborious absorption tests

became the basis for the serological differentiation of the Salmonella for over twenty years and our present-day knowledge of the fundamental relationships within the group was built up in this way. By means of these tedious cross-absorption tests the Salmonella group began to be classified, attempts such as those of W. G. Savage and H. Schutze forming the basis of the extensive work of P. B. White in the early 1920s. White was employed by the Medical Research Council to collaborate with W. G. Savage, the medical officer of health for Somersetshire, in the investigation of outbreaks of food-poisoning. He worked first at Bristol and later at the Lister Institute, becoming very interested in the taxonomy, classification and interrelation of the various species of Salmonella. The results of his most laborious researches were published, in Medical Research Council reports, in 1925 and 1926. He distinguished many different species of Salmonella, arranged them in groups and suggested antigenic formulae for them. His scheme of antigenic notation, however, lacked the clarity of that propounded some years later by F. Kauffman, and, at the meeting of the international sub-committee on Salmonella nomenclature, in 1934, a taxonomic scheme based on that of Kauffman which was in fact 'a re-examination and amplification of the pioneer work of Schutze, Bruce White, Scott and others' was adopted, and is known today as the Kauffman-White scheme.

F. W. Andrews made two important contributions to the identification of Salmonella species. He introduced the use of mono-specific antisera, that is, antisera which had been absorbed in bulk with heterologous organisms and which were thus specific for a given species. The use of mono-specific sera much simplified the identification of related species, eliminating the necessity for cross-adsorption tests. The use of mono-specific antisera led Andrews to observe that all the organisms in a given culture might not be agglutinable by the specific antiserum and thus to the concept of Group and Specific phase variation.

For convenience, our account of the Salmonella has up to this point been anthropocentric but it was early appreciated that organisms of this group could behave as animal pathogens. In fact the first Salmonella to be isolated, except the typhoid bacillus itself, was *Salmonella cholerae-suis* which D. E. Salmon and his pupil Theobald Smith isolated from pigs suffering from

hog-cholera. In 1889 F. Leoffler isolated the organism now known as *Salmonella typhi-murium* from a colony of laboratory mice suffering from an epidemic disease. This organism was later shown to be the same as the Aertryck bacillus. However, the generic name 'Salmonella', proposed in 1900, has become accepted and honours the first man to isolate a member of this genus although, in fact, he was a minor student of this group of organisms,

Knowledge of the Enterobacteriaceae and of bacteriology in general was further extended by the investigation of practical problems arising from the use of the agglutination reactions in the diagnosis of typhoid fever, particularly during and after the First World War.

Until the First World War some form of the microscopic technique for the agglutination reaction was usually employed, even though there were a number of disadvantages. For instance, since the observation of loss of motility formed part of the test it was necessary to use a living broth culture which was both dangerous and difficult to standardize. The variability inherent in the method made it unsuitable for the quantitative measurement of the amount of agglutinin in the serum, that from any one patient giving irregularly different figures from day to day. In 1906 G. Dreyer, then working at the Staatens Serum Institut, Copenhagen, introduced his technique of making serial dilutions of serum, in small test-tubes, to which were added standard suspensions of selected, sensitive, killed organisms, the results being read macroscopically. This technique was published in English in 1909, but it was not until the First World War that it superseded the microscopic technique. The importance at that time of being able to make a diagnosis in patients with typhoid fever who had received injections of T.A.B. vaccine made essential an accurate quantitative technique which could reliably determine a rise in titre of agglutinin as the disease progressed. Dreyer had been appointed Professor of Pathology at Oxford in 1907 and, during the First World War, he and his colleagues devoted much attention to the technique of serological diagnosis in enteric fever. They defined its use and limitations with a thoroughness not achieved before and issued standardized suspensions of organisms for use in the armed forces.

The confusion in the serodiagnosis of enteric fever by the techniques in use during the First World War was so great that serologists of vast experience were of the opinion that 'the Grüber-Widal reaction has lost its practical value in consequence of the antityphoidal inoculation'. However, the investigation of this difficulty not only resulted in a clarification of the situation, with regard to the diagnostic Widal test and its restoration to a place of importance in the diagnosis of enteric fever, but also to one of the most fruitful lines of bacteriological investigation ever undertaken.

Prominent amongst the investigators of the agglutination reaction was A. Felix (1887–1956). Like P. B. White (1891–1949) he was not medically qualified and came to bacteriology as a form of war work during the First World War. Working with Weil (1848–1916) on the aetiology of typhus fever, he discovered the remarkable fact that the serum of typhus patients contained an agglutinin against certain strains of Proteus, although this organism played no part in the causation of the disease. Moreover, it was noted that although the serum of typhus patients might agglutinate the special strain of Proteus designated X 19, to high titre and ordinary Proteus strains not at all, a rabbit antiserum prepared against Proteus X 19 agglutinated all strains of Proteus equally. The character of the masses of agglutinated bacteria was also different, being coarse and fluffy with the rabbit antiserum and fine and granular with the typhus serum. Weil and Felix soon showed that Proteus X strains might exist in two forms distinguishable by colonial appearance and designated H and O forms. The explanation of the difference between rabbit antiserum and typhus serum seemed to be that rabbit serum contained antibody against both the H and O types but the typhus serum only against the O.

In 1918 they showed that the same applied to Salmonella paratyphi B. After the war Felix went to Israel as director of a bacteriological laboratory in Tel Aviv. Here was a vast supply of material for study in the routine Widal tests, some of which were on serum from inoculated and some from uninoculated persons. As a result of an intensive study of the Widal reaction in these two classes of cases Felix developed the technique of 'qualitative receptor analysis' and, in a classic paper published in 1924, came to the following highly important conclusions:

(1) That antityphoid inoculation produced only antibody causing coarse aggregation of the bacilli, 'large flaking antibody', and thus a diagnosis of typhoid fever in an inoculated person could be made if only 'small flaking' antibody was looked for.

(2) Although actual typhoid infection caused the production of both types of antibody there was no correlation between the amount of 'large flaking' antibody and the clinical course of the disease, whereas the early production of 'small flaking' antibody was a good prognostic sign. (As a result of these observations he was led to question the value of antityphoid inoculations and to his later work on the special Vi strains of typhoid bacilli for vaccine production.)

(3) That all antisera to organisms of the enteric group contained two types of antibody, the 'large flaking' which corresponded to a labile antigen on the organism, and the 'small flaking' which corresponded to a stable antigen. The 'small flaking' antibody showed cross-reaction between different members of the enteric group but the 'large flaking' antibody was specific.

Felix used as antigens to detect these two types of antibody specially selected strains of live organisms, one of which would only react with the 'small flaking' antibody. Although other workers showed that heated or alcohol-treated organisms ceased to react with the 'large flaking' antibody but continued to be agglutinated by the 'small flaking' antibody, Felix always preferred selected live organisms for the Widal test. He recommended, for the diagnosis of enteric fever, that the patient's serum be put up in dilutions of 1/100 and 1/200 only with B. typhosus (H and O variants), B. paratyphosus B (H strain), B. aertrytke (O variant because this had the same O antigen as B. paratyphosus B), and two different strains of B. paratyphosus A, one of which was sensitive to both H and O antibodies and one which was sensitive to H antibody only, no pure 'O' variant of this organism being known.

A serological tool of the greatest value which was brought into practical medical bacteriology in the early years of this century was the complement fixation reaction. The discovery of the principles underlying this type of test by Bordet has already been described. But Bordet was also the first to appreciate that

these observations offered a method for the detection of antibody or antigen in cases where other techniques were inapplicable. Using this method he demonstrated antibody to avian tubercle bacilli in the serum of an infected guinea-pig. It was soon shown by various workers that absorption of complement occurred whenever antigen and antibody combined irrespective of the nature of the antigen. It was found that the test could be adapted to detect antigen, using a known antiserum. The test was extraordinarily delicate and soon had an important material application to the detection of blood in medicolegal work. Probably the first practical diagnostic use of the complement fixation test was made by Widal who used the test to detect typhoid antibodies in patients. He found that the test became positive earlier than the agglutination reaction. He was also the first to explore the possibility of this test being an aid to the diagnosis of tuberculosis but it was soon discovered that the test was not positive in a sufficiently large percentage of patients to be of great value.

Wassermann and Sachs, using the same technique, tested tuberculous organs for tubercle antigens, using known antituberculous sera and obtained positive results in some cases. The importance of these experiments lies in the fact they suggested to Wassermann and Sachs the use of syphilitic organs as antigen when the causative organism, the *Treponema pallidum*, was discovered in 1905. Their first results in this field, published in 1906, showed that the sera of monkeys infected with syphilis gave a positive complement fixation test with extracts of syphilitic lesions of man or monkeys. Just fourteen days after the publication of Wassermann's paper, Detre published the results of complement fixation tests on the sera of six syphilitic and four normal persons using syphilitic organs as antigen. He obtained positive results in two of the syphilitics. Wassermann and his colleagues studied the C.S.F. of patient with G.P.I. and found a high proportion positive but, on the other hand, in a large series of syphilitic cases of all stages of the disease, seen in Neisser's clinic, only 19 per cent of cases were found to give a positive result with this test. It was not until later work had shown the importance of the quantitative aspect of the test that more consistent results were obtained.

The immediate, relative failure of the complement fixation

test as a diagnostic aid in syphilis led Neisser and Bruck to revert to the idea of detecting syphilitic antigens in the blood. For this purpose they prepared extracts of the erythrocytes of syphilitic patients and used the serum of highly immunized monkeys as a source of antibody. This appeared to work well but it was almost immediately discovered that extracts of normal erythrocytes gave similar results. These results were followed by the report of a positive reaction with syphilitic serum using an extract of human spleen as antigen and by the work of Landsteiner who showed that an alcoholic extract of guinea-pig heart also reacted with syphilitic serum to fixation of complement.

These results dealt a death blow to the theory of the Wassermann reaction and, when taken in conjunction with the widespread use of the test by inexperienced persons and its many inherent sources of error, they tended to retard the appreciation of the value of the test; it required several years of careful quantitative work to establish its sphere of usefulness.

The complement fixation test was applied as a diagnostic aid to a number of other infections such as gonorrhoea, tuberculosis and pertussis, but, although positive results were obtained with each of these diseases, the test did not prove valuable in practice except in respect to gonorrhoea. In this condition, particularly in its chronic form and when once a good technique of antigen preparation had been developed, it proved useful.[197]

Even when a particular bacterium had been shown to be the cause of an infectious disease it was often years before the elementary aspects of the epidemiology of the infection were elucidated. A good example of the gradual understanding of the epidemiology of a disease, long after the causative organism had been discovered, is the condition now generally known as brucellosis.

We have already described how Bruce discovered the cause of Malta fever in 1887, but how the organism spread in the community and gained access to the human body remained unknown for almost twenty years. In 1897 another British army medical officer, M. L. Hughes, after a six-year tour of duty in Malta, published a comprehensive monograph on, *Mediterranean, Malta, or Undulant Fever*, based on his own extensive experience. Hughes fully confirmed Bruce's bacteriological

findings and his discussion as to the mode of propagation and dissemination of Malta fever must now be considered. He pointed out that there was no evidence that the disease was contagious nor that the causative organism gained access to the body through the skin via wounds or mosquito bites. By a process of exclusion, therefore, it must enter via the alimentary canal or air passages. Hughes considered the possible relevance of a polluted water supply in great detail and concluded that this was certainly not relevant. Likewise, he thought food was not likely to be a vehicle, since the food of soldiers was of a standard quality and prepared under close supervision. He dismissed the possibility of milk playing a part on the negative grounds that he had 'met with no fact that would favour a causal connection between milk supply and this fever' and the fact that he had 'known undulant fever to attack families who used only Swiss condensed milk, regiments in which no other milk was allowed in barracks, and families whose milk supply was always milked from goats on their own premises, into their own vessels, under reliable supervision'. The inaccuracies inherent in the first two parts of this statement illustrate the difficulty even the most careful observer has in the collection of reliable epidemiological data. Hughes considered that, in his experience, undulant fever was associated with insanitary surroundings particularly 'some insanitary condition connected either directly with drainage or with flooring of an absorbant nature which has had probabilities of contamination with human excrement'. He described no less than fifteen outbreaks which appeared to substantiate this hypothesis. Although aware that microbes were not generally numerous in the air, Hughes considered that the causative organism spread from faecal contaminated material through the air and entered the respiratory passages.[198]

In 1904, so great was the problem of Malta fever among British troops stationed in that island, the government arranged for a commission to study the epidemiology of the disease on the spot. Partly by luck, the essential role of the goat and their milk was discovered almost immediately by W. H. Horrocks and T. Zammit. These workers wished to know what animals could be infected experimentally with Micrococcus melitensis and decided to try goats which were readily available. Six goats were

bought cheaply and, as a preliminary to their inoculation, Zammit tested their blood for agglutinins against Micrococcus melitensis. To their surprise five were positive. Culture of the goat's milk yielded an abundant growth of the organism. Further work soon showed that a high proportion of goats on Malta were infected and that pasteurization was an effective means of killing the organism. The essential epidemiology of Malta fever was thus elucidated.[199]

Meanwhile an apparently quite unconnected discovery had been made by the Danish doctor and veterinarian B. Bang, in 1897. He investigated the cause of epizootic abortion in cattle from a bacteriological point of view. He was not the first so to do, having been antedated by Nocard who isolated a variety of organisms from the uteri and placentas of aborting cows. Bang's technique, modelled on the approach of Koch's school, was vastly superior to Nocard's and almost at once yielded significant results. He appreciated that he must examine a cow, showing signs of impending abortion, before abortion took place and that the uterus and contents must be removed and examined rapidly with aseptic precautions. Microscopy of the fluid exudate between the uterus mucosa and the foetal membranes, showed the presence of a very small bacterium, apparently in pure culture. Bang attempted to culture the organism and was particularly fortunate in his choice of method; he inoculated melted serum-agar-gelatine in tubes which were allowed to set upright. He found that the bacterium grew well but only in a zone a few millimeters below the surface of the medium – neither in the more aerobic medium above nor in the more anaerobic medium below. This unusual characteristic was of course of great diagnostic value. Bang rapidly showed that this organism was to be found in a high proportion of cases of epizootic abortion and rounded off his work by showing that inoculation of a healthy pregnant cow with a pure culture of the organism resulted in abortion.[200]

The further development of our knowledge of brucellosis came largely from workers in the United States. Workers at the Bureau of Animal Industry, including Theobald Smith, showed that Bang's bacillus occurred among cattle in the United States and that it was excreted in the milk. Smith and his colleague, M. Fabyan, therefore, suggested, in 1912, that it would be worth

investigating the possibility that Bang's bacillus might be the cause of some disease or other in man.[201] Taking up this point, in the following year, W. P. Larson and J. P. Sedgwick, showed that the serum of 73 out of 425 children contained antibody to Bang's bacillus but they did not connect infection with any particular clinical syndrome.[202]

The next advance came from the work of Alice Evans who started with the thought that since the organisms causing Malta fever, and contagious abortion were both excreted in milk it might be worth comparing them in detail. She showed that the 'Micrococcus melitensis' was in fact a short bacillus and both culturally and serologically the two organisms were closely related. Both organisms were agglutinated by a serum prepared against either and they could only be distinguished by the fact that the homologous organism was agglutinated to higher titre.

Alice Evans's work on the Brucella group of organisms (as they became to be called about this time) extended from 1917 onwards but, at the time she established the connection between the 'Micrococcus melitensis' and 'Bacterium abortus', there was no suggestion that the latter caused undulant fever in man.[203] The first reports of such cases were made by L. E. W. Bevan in Southern Rhodesia.[204] The first clear and thoroughly investigated case was reported by C. S. Keefer of Baltimore in 1924. His patient who had become infected from cow's milk, ran a typical undulating fever and an organism characterized as Brucella abortus was isolated from his blood on several occasions.[205] Gradually the number of reported cases of human disease due to Brucella abortus increased although it seemed largely true that, in countries where goat's milk was not consumed, undulant fever, in its typical form, hardly appeared to occur. With increasing experience and more careful diagnosis of febrile disorders this paradoxical situation was gradually seen to be more apparent than real, but it is only in recent years that the extent of the problem of brucellosis in England has been fully appreciated.

The new knowledge of bacteria allowed the rationalization of some traditional public health practices. Prevention of infectious disease in the community was, even in the first decade of the twentieth century, largely a matter of tradition. The idea that the causative organisms of disease generally multiplied in

filth and thence spread through the air died hard. Contact with infected patients was assumed to be a major hazard in most diseases and spread by families of great importance. But precise epidemiological knowledge for particular infections was largely lacking.

One of the foremost of the new school of public health workers, who tried to adapt their sanitary practice to the new facts of bacteriology, was C. V. Chapin, superintendent of health in Providence, Rhode Island. His book, *The Sources and Modes of Infection*, published in 1910, has become a classic of the public health literature. Chapin sought to base his preventive measures, as far as possible, on exact knowledge, bacteriological and epidemiological, of the sources and methods of spread of individual infective agents. In particular he endeavoured to assess the quantitative importance of different methods of infection. He wrote: 'We need to measure more carefully the relative importance of different sources of disease and different modes of infection. It is not so important to know that typhoid bacilli live in water for weeks, as it is to know that 99 per cent die in one week. It is not enough to discover that diphtheria bacilli can be recovered from articles in the sickroom, we must learn how often they are found and how often disease is traced to such a source. . . . Doubtless the house-fly has been the cause of typhoid fever, but in which percentage of cases we are profoundly ignorant.'

One of the most convincing examples of the value of the systematic application of bacteriological methods to public health was the anti-typhoid campaign conducted by the Prussian government in south-west Germany. Under the general direction of Robert Koch, an organization of eleven 'bacteriological stations' were set up, each with a director and from one to three assistants as well as laboratory attendants. The task of these centres was to assist practitioners with accurate bacteriological diagnosis of typhoid fever, to try to trace the source of infection and to make sure that hygienic precautions were not relaxed until known cases of the disease had recovered and ceased to excrete typhoid bacilli. As a result of this work a number of important new points, clinical, epidemiological and pathological regarding typhoid fever soon emerged. It was found that the number of cases of typhoid fever actually notified

by practitioners grossly underestimated the number of cases of the disease. If, as a result of notification on clinical grounds, the population around the index case was investigated bacteriologically many more cases of typhoid would be found, ill but with an atypical clinical picture. The importance of the gallbladder as a site of multiplication of the typhoid bacilli, not the intestine as had been supposed, was demonstrated and, above all, the importance and frequency of chronic carriers of the typhoid bacillus was established. In addition a number of technical advances concerning the isolation of typhoid bacilli from clinical specimens were made. Based on the work of the bacteriological stations, improvements were made in the sanitation and water supplies in endemic areas, so that the number of cases of typhoid fever in the area served by the bacteriological stations fell from 3,487 in 1904 to 1,226 in 1909 and permitted the hope that 'in no far distant time the civilized world may be freed from this scourge of mankind'.[206]

From the public health point of view the most important contribution of bacteriological science was the demonstration of the importance of carriers and 'missed cases' in the spread of infectious disease. Although it had long been appreciated that mild cases of infectious diseases, difficult or impossible to recognize, did occur, they were regarded as rare and unimportant. Bacteriological studies showed that in some infections atypical cases were very common and that completely healthy 'carriers' of the causative organism might be common. It was Chapin, particularly, who drew attention to the importance of such cases from a public health point of view. He marshalled evidence showing that in such diseases as typhoid fever, cholera, diphtheria and meningococcal meningitis, not only did carriers occur, but that particular outbreaks of infection could be traced to such sources and stressed that any scheme of prevention which failed to take account of carriers and missed cases was doomed to partial or perhaps complete failure. In Great Britain J. C. G. Ledingham and J. A. Arkwright were the chief apostles of the 'carrier' doctrine and, in 1912, published their well-known monograph on 'the carrier problem in infectious diseases'. As techniques for distinguishing different 'strains' of organisms within a single species were developed the importance of carriers in certain infections was confirmed and

extended. The painstaking investigations of Dora Colebrook into the epidemiology of puerperal fever due to the haemolytic streptococcus is a good example. Using specific agglutinating antisera, supplied by F. Griffith, she was able to demonstrate a potential source of infection in a high proportion of cases. Most important was the fact that in half the cases of this highly lethal disease the source appeared to be one of the patients' medical attendants.[207]

A discovery which was to prove of the greatest value in the epidemiological study of bacterial infections was that of F. W. Twort who, in 1915, described the agent which later became known as 'bacteriophage'. Twort, the son of a doctor practising in Camberley, was born in 1877 and, like the rest of his family, studied medicine at St Thomas's Hospital Medical School where he qualified in 1900. His first job was as assistant to Louis Jenner, the director of the clinical laboratory but, in 1902, he became assistant to W. Bulloch at the London Hospital where he remained for seven years and was responsible for the whole of the diagnostic bacteriology. In 1909 he was made superintendent of the Brown Institute and remained so until it was destroyed by a bomb in 1944. Twort was a straightforward, shy individual with a slightly paranoid disposition but he had a marked streak of originality. Not for him the universal preoccupation of his contemporaries with vaccines and the opsonic index; he struck out on his own lines, largely unprofitable though they were, on the origins of life and the evolution of pathogenic microbes from free-living organisms. Twort stumbled across the effects of bacteriophage when engaged in the search for non-pathogenic ultra-microscopic viruses which he believed must exist in the same way as did nonpathogenic bacteria. During the course of culturing calf lymph on agar he isolated a micrococcus which exhibited unusual properties – it tended to develop glassy, transparent areas and microscopic examination of such areas showed that the micrococci had disappeared. He showed that if some of the culture from a glassy area was emulsified in water and filtered through the finest porcelain filters the filtrate contained an agent which was lethal to the micrococcus. He found that this 'disease of the micrococcus' could be transmitted to fresh micrococci 'for an infinite number of generations'. Twort discussed the various possibilities as to

the nature of this agent without coming to any definite conclusions, but he did raise the point that it might be some sort of ultra-microscopic virus capable of growing in bacteria, indeed one of the non-pathogenic viruses for which he was searching.[208]

Although Twort was the first to describe, although not name, the phenomena of bacteriophage activity we owe our basic knowledge of the subject largely to the French-Canadian Felix d'Herelle. He was born in 1873 and, although retaining his British nationality all his life, was more French than British in temperament and outlook. He went to school in Paris but studied medicine in Montreal and few people can have travelled more extensively in the pursuit of their profession. d'Herelle was professor of bacteriology in Guatemala from 1901 to 1907, bacteriologist to the Mexican government, worked in Argentine, Turkey, Tunis, Egypt, Indo-China, India and Holland and, between 1934 and 1936, was entrusted by the government of the U.S.S.R. with the setting up of bacteriophage laboratories at Tifles, Kiev and Kharkov. He was also on the staff of the Pasteur Institute in Paris, where his most important work was done, and for five years was professor of protobiology at Yale. He died in 1949.[209]

In 1917 d'Herelle published a short but most interesting paper entitled '*Sur une microbe invisible antagoniste des bacillus dysenteriques*' in which he described how he had discovered, in the faeces of patients with dysentery, a filterable agent which, when added to a broth culture of Shiga's dysentery bacillus caused the death by lysis of that organism. He showed that this agent was transmissible from generation to generation of bacillus and clearly postulated, as the title of his paper indicates, that he was dealing with a parasite of the dysentery bacillus. d'Herelle also showed that his agent could be adapted to produce lysis of Flexner's dysentery bacillus but not the typhoid bacillus or staphylococci. But, working with the faeces of a patient convalescent from enteric fever, he isolated a similar agent active against the paratyphoid bacillus.[210] d'Herelle coined the term 'bacteriophage' and made the subject very much his own. A succession of papers on the subject culminated in a 277-page monograph in 1921 and, in 1926, he published a book of over 600 pages entitled *The Bacteriophage and Its*

Behaviour. There is no doubt whatever that d'Herelle made the most important contribution to the discovery and elucidation of bacteriophage and its effects but, unhappily, his nature contained a measure of meanness and petty jealousy which made him unwilling to allow the smallest share of credit to any other worker. He was willing to allow that a number of workers might have observed bacteriolytic phenomena, perhaps due to phage, but since they had neither investigated it nor grasped its essential nature, they were certainly entitled to no credit. However, Twort, as we have seen, had described bacteriophage and its action and its essential character of a self-replicating agent which destroyed bacteria, not quite as clearly as had d'Herelle, but in a perfectly intelligible manner almost two years earlier. d'Herelle, however, tried strenuously to dissociate Twort from bacteriophage by, firstly, maintaining that the phenomenon observed by Twort on a solid medium was quite distinct from the phenomenon he had described (he obligingly coined a special name for Twort's phenomenon – 'bacterioclysis') and, secondly, by trying to antedate his own observations on bacteriophage to the period between 1911 and 1914, when he was investigating a diarrhoeal disease of locusts in Central America. The former line of argument was quite unconvincing, and indeed displays d'Herelle in a rather ridiculous light and his second contention was simply dishonest.[211, 212, 213] d'Herelle's chief interest in bacteriophage, apart from its fundamental character, was in its clinical significance. He investigated the bacteriophage content of the faeces of patients with bacillary dysentery and typhoid fever and concluded that the rate of recovery of the patient was largely dependent upon the speed with which the bacteriophage became adapted to the invading parasite and destroyed it. It was natural, therefore, that attempts should be made to use bacteriophage therapeutically. After some trials he was convinced that bacteriophage provided a specific therapy in dysentery and staphylococcal infections and that preliminary results in plague and other infectious diseases were encouraging.

It was the work of J. Craigie and his colleagues to draw attention to what has proved the most important practical aspect of bacteriophage – its use as an epidemiological tool to distinguish different strains of bacteria within a single species.

Before describing Craigie's work a brief digression must be made on a new development in our knowledge of the antigenic structure of the typhoid bacillus, published by A. Felix and Margaret Pitt, in 1934. They started from the observation that strains of typhoid bacillis which were highly sensitive to agglutination by O somatic antisera were also readily killed by normal human serum *in vitro*. Some strains of typhoid bacillus were known not to be readily agglutinable by O antisera and, therefore, ought to be resistant to the bactericidal action of serum and more virulent than O agglutinable strains. Felix and Pitt decided to test this assumption using the mouse as their experimental animal. They found, as expected, that O inagglutinable strains of the typhoid bacillus were more virulent to mice than O agglutinable strains. Investigating the matter further they found that O inagglutinability was associated with the presence of a hitherto undescribed antigen which, because of its correlation with mouse virulence, they called the 'Vi' antigen. They consider the Vi antigen of the greatest importance since immunization of mice with a vaccine made from a Vi positive strain of the typhoid bacillus gave them good protection against subsequent challenge, but a vaccine prepared from a Vi negative strain gave insignificant protection. The most important implication for man was, in their opinion, that strains used for the manufacture of vaccines must have the highest possible content of Vi antigen. They remarked that 'Doubts will possibly be voiced as to whether the facts established in experiments with mice are also valid with regard to the disease in man. Such a view, however, does not seem to carry weight'. [214, 215] In fact, from the practical point of prophylactic immunization in man, these observations turned out to be a classic example of the dangers of extrapolating from mice to men for, when a field trial was made, using a vaccine in which the Vi antigen had been carefully preserved, it was found to be far less effective than the old-fashioned heat-killed suspension of typhoid bacillus as originally used by Wright.

But to return to the work of J. Craigie; he and C. H. Yen noted that bacteriophages, or 'phages' as they had become to be named, existed which would only attack strains of the typhoid bacillus containing Felix Vi antigen. In fact, further work showed that there were four phages with this characteristic, but,

which differed in other respects such as particle-size, the temperature required to inactivate them and in antigenic structure. They, therefore, attempted to divide up Vi positive typhoid bacilli into strains based upon their reactions with these four phages, but the results proved unsatisfactory. In particular there were many discrepant reactions with the phage they designated 'type II'. Investigating these discrepancies Craigie and Yen showed that the specificity of type II phage depended upon the strain of typhoid bacillus upon which it had been propagated – the phage developed the power to lyse the propagating strain but not others. They were thus able to prepare a series of phages specifically adapted to different strains of typhoid bacillus and, when testing these phages against groups of typhoid bacilli of known origin, the phage reactions were consistent with the epidemiological facts – in clearly defined outbreaks traced to a particular source all the strains of typhoid bacilli isolated, including the supposed source, reacted with the same phage.[216] Craigie and Yen developed a simple, practical typing scheme and established the value of the principle of phage-typing for epidemiological purposes which has been subsequently adapted to the typing of a number of other species of bacteria, most notably the *Staphylococcus aureus*.

8 The Chemotherapy of Bacterial Disease

Scientifically based chemotherapy was invented by Paul Ehrlich and may be said to have resulted from the imaginative application of his interest in the chemistry of living organisms which first clearly showed itself whilst he was still a medical student. Ehrlich himself recalled that his interest was aroused when, as a student, he read a paper on the distribution of lead in animal tissues and at that time, 1877, he lost a few months in a fruitless attempt to follow this up experimentally. Many years later, in a letter to C. A. Herter, Ehrlich wrote, 'I also in fact believe that my real natural endowment lies in the domain of chemistry; and mine is indeed as you know, a kind of visual, three-dimensional chemistry. The benzene rings and structural formulae really disport themselves in space before my mind's eye, and I believe that it is just this faculty which has been of supreme value to me in my later studies.' From the very first Ehrlich had the idea that different chemical substances had differing affinities for various living cells and their contents and, after his initial fruitless efforts, turned to research with the recently discovered analine dyes. As early as 1879 he had demonstrated the general validity of his doctrine of specific affinities, in respect to the staining of the blood leucocytes, in his classic paper *'Uber die specifischen granulationen des blutes'*. In 1881 he introduced the dye methylene blue into bacteriology and, the following year, within a few weeks of Koch's announcement of the discovery of the tubercle bacillus, such was his grasp of the principles involved in specific staining reactions, Ehrlich was able to devise, in principle, the technique of acid-fast staining which is still used today. In 1885 he published his thesis for recognition as a university teacher entitled 'The requirement of the organism for oxygen – an analytical study with the aid of dyes' in which, using the dyes alizarin blue and

indophenol blue, the colour of which varied according to the oxidation-reduction potential of their environment, he endeavoured to investigate the vital combustion process. For over ten years after qualification Ehrlich's researches were largely devoted to pharmacological problems, in particular, the affinity of methylene blue for the nervous system, its possible use as an anaesthetic, the possible changes in its structure which might make it more effective and the value of cocaine as a local anaesthetic. Quite early in his career Ehrlich had developed the concept that specific chemotherapy of disease due to parasites of all kinds was possible. Following out his ideas on the specific affinities of chemical substances for particular living cells he developed the hypothesis that substances could be produced which, whilst having great affinity and toxicity for an infecting parasite, would not react with host tissues and would therefore be harmless to the host.[217] As early as 1891 he was able to demonstrate the truth of his concept in the case of methylene blue as a chemotherapeutic substance in malaria. It had been known for some time that methylene blue was a useful stain for malaria parasites and that if a drop of malarious blood was mixed with dilute methylene blue, so that the blood cells and plasma were scarcely tinged blue, the malaria parasites were intensely strained. Ehrlich was able to demonstrate that the administration of methylene blue to two cases of malaria, seen in Berlin, resulted in rapid clinical cure.[218] He also saw that potentially useful chemotherapeutic substances might have their chemical constitution so altered as to have more of the desired effect and less of the undesirable host toxicity. This he described as 'learning to shoot (to take aim) and do so through chemical variations'.[219] However in a series of investigations, about 1890, in which he attempted to treat experimental bacterial infections in vivo with synthetic antiseptics he was entirely unsuccessful and did not even find a promising clue as to the direction in which a further search for chemotherapeutic substances might take.

Thus he could not but be impressed by Behring's discovery of antitoxins in 1891, for these substances seemed to have, to such a high degree, the properties of 'charmed bullets' for which he was seeking in chemotherapeutic substances. The following decade, which was devoted to his great work in immunology,

was, in many ways, but a diversion from his main line of interest in practical chemotherapy.

Ehrlich's active return to the field of chemotherapy, about the year 1902, was conditioned by two main factors. The first must have been that his purely immunological work had become 'bogged down' in a series of sterile arguments and in his attempts to fit every new immunological observation into the framework of his 'side-chain theory'. The second factor was the sudden availability of a very convenient experimental model. In 1902 Laveran had reported that the trypanosome which causes mal de caderas in horses was transmissible to rats and mice – highly convenient laboratory animals. Moreover the disease thus produced ran a uniform, rapidly fatal course and the trypanosomes were readily visible, in large numbers, in the blood of the animals. In addition Ehrlich considered that the larger, more highly developed trypanosome would be a more susceptible target for chemotherapeutic agents.

In a remarkably short time Ehrlich discovered, following his own theoretical principles, but based on what can only be described as intuition, an active chemotherapeutic substance. He described just how, in the letter to C. A. Herter, previously quoted. Harking back to his dyes, he wrote, 'It had always struck me that the benzopurpurins remain so extraordinarily long in the body, and I had always said: "This class of substances must be able to do something or other of a special kind." Every few years I had the stuff subjected to a new test, and when I came into possession of the trypanosomes, benzopurpurin was the first substance which I had tested on them. There was evidence, when it was applied in the test, of an activity which, though quite unmistakable was slight; and this I attributed to the poor solubility of the substance. I accordingly asked Weinberg (his organic chemist) to have it made rather more soluble by the introduction of a sulphonic acid group and the thing was done – trypan red was discovered.' Trypan red proved to have a remarkable effect in the treatment of mice infected with the trypanosome of mal de caderas and fully established the validity of Ehrlich's general line of approach to chemotherapy. However it was much less use in infections with other trypanosomes and proved of little use in

clinical practice. But, almost at once, another promising chemotherapeutic substance came to hand – atoxyl.

Atoxyl is an organic arsenical which had been synthesized some thirty years before and had had a limited vogue in medicine as an alleged treatment for anaemia. Now there was good empirical evidence that arsenic in its simple, inorganic form had some trypanocidal action. Lingard, in India, had tried, on a quite empirical basis, a large number of chemicals in the treatment of surra in horses. None had had any effect except arsenic. One out of twenty-three horses treated with large doses of arsenious acid survived and was in good health over three years later.[220] Bruce had also tried inorganic arsenic in cattle with nagana, with some beneficial effect, and this had been confirmed by a district medical officer in Nigeria, E. J. Moore, in 1904. Further, Laveran had shown that sodium arseniate had a trypanocidal effect in experimentally infected mice but its toxicity made it relatively useless in practice.

It was H. W. Thomas, a graduate of McGill University, who was working at the Liverpool School of Tropical Medicine, who drew attention to the potential value of atoxyl in the treatment of trypanosomiasis. Thomas had done important work showing that trypanosomes from cases of human sleeping sickness from Uganda, the Congo and the Gambia behaved in the same way in experimental animals and were, therefore, probably the same species.[221] He showed in 1905, that atoxyl, which contained nearly 38 per cent of arsenic yet was almost non-toxic, was an effective trypanocidal drug in experimentally infected animals.[222] About five hours after the administration of atoxyl the trypanosomes could be seen to be undergoing dissolution in blood samples from the infected animals. Although sufficiently promising to warrant trial in human sleeping sickness, atoxyl was by no means a 100 per cent cure.

It was at this point that Ehrlich's genius and knowledge of organic chemistry were displayed. The accepted structural

formula for atoxyl was With the arsenical radicle

inserted into the amino group, atoxyl was supposed to be chemically rather inert and not capable of much modification. Ehrlich

insisted, in the face of expert chemical opinion (and two of his three senior organic chemists are said to have resigned over the issue) that the arsenic radicle was inserted directly into the

benzene ring giving the following structural formula

The amino group was thus left freely available for the insertion of a wide range of other chemical groups. Ehrlich and his remaining chemist, Bertheim, were able to prove that their concept of the structure of atoxyl was correct and set about producing a great range of variations of the atoxyl molecule which were then tested for their curative effects in mice and rats infected with trypanosomes, It was this particular idea that Ehrlich regarded as his most original contribution to practical chemotherapy and which led directly to an effective treatment for the human trepanematoses. The 418th compound tested, arsenophenylglycin, was very effective; the 606th, later to be so famous as the cure for syphilis, was not regarded as promising and put on one side.

In 1905 F. Schaudinn and E. Hoffmann discovered the causative organism of syphilis, *Treponema pallidum*. And Schaudinn also put forward the view that the spirochaetes, of which *Treponema pallidum* was one, were closely related to the trypanosomes. To this day the evolutionary relationships of the spirochaetes are very obscure and any relationship to the flagellates as a whole, let alone the trypanosomes, must be distant indeed. None the less Ehrlich accepted this idea and adduced additional clinical evidence in its favour. He pointed out that the degenerative changes produced in the central nervous system of dogs by trypanosomiasis was not unlike that produced in syphilitic tabes and that one trypanosomal disease of horses, dourine, was sexually transmitted. He was not alone in this view and himself denied any originality here. Professor Lassar and his colleagues, who were known personally to Ehrlich, had, in 1907, used atoxyl in the treatment of syphilis on the same theoretical basis.[223] When therefore Ehrlich's old friend, Professor Kitasato, sent one of his pupils, S. Hata, who had already, in Japan, developed a technique for infecting rabbits with the spirochaete of syphilis, to Ehrlich, for post-

graduate work, it was natural that he should be put to work testing Ehrlich's large accumulation of atoxyl derivatives for antisyphilitic activity. It was thus that the value of '606', arsphenamine, as a treatment for syphilis was discovered.[224] There was nothing new in the use of an arsenical preparation in the treatment of syphilis. It had been suggested as long ago as 1832 and throughout the nineteenth century, in various forms, it had been used. Arsenical preparations were not so popular nor probably as effective as the older remedies, mercury and potassium iodide, but they were definitely thought clinically to be of value. Atoxyl was used with apparent success in the early 1900s.[225]

Arsphenamine was first used in man, not to treat syphilis but to treat relapsing fever. J. Iverson in St Petersburg, in 1909, had shown that arsenical preparations such as atoxyl had a striking effect on this spirochaetal disease which was a very convenient test condition with its characteristic relapsing fever and the easily seen spirochaetes in the blood. Ehrlich therefore suggested that Iverson might like to try '606'. During the usual winter epidemic of relapsing fever in St Petersburg Iverson was able to test the new drug in fifty-five patients. Fifty-one were completely cured by a single injection. As '606' had never been tried in man before Iverson had to proceed very cautiously and gradually worked out the technique, that was eventually used universally, of administering the drug in an intravenous drip. Ehrlich had recommended subcutaneous or intramuscular injections but these produced long-lasting and very painful local reactions.[226] The first reports of the value of '606' in the treatment of human syphilis began to be published in 1910 and it was soon apparent that a remarkable new therapeutic substance had become available. Almost within hours of administration some of the more acute manifestations of syphilis, such as ulcers in the mouth and painful periostitis, were dramatically improved and benefit was also noted in the more chronic forms of the disease. Initially it really seemed that Ehrlich had achieved his ambition to produce a chemotherapeutic agent, one injection of which, would completely cure the patient.[227]

Thus was the first really useful chemotherapeutic substance introduced into medicine. It is not easy to give a succinct

account of so large and important series of observations but we believe that the story as just related is an accurate summary of the main steps. Ehrlich himself said that successful medical research required the four big 'Gs' – Geduld, Geschick, Geld and Gluck (patience, ability, money and luck) and it is of interest to try to separate the various components in the relation to the discovery of '606', the greatest achievement of Ehrlich's most productive life. In this discovery Ehrlich's intuitive genius played by far the largest role and may be broken down into three main parts. First, the simple concept of chemotherapy, that it was possible deliberately to synthesize substances which by nature of their chemical structure were toxic to the infecting parasite but harmless to the host, and demonstrating the validity of the idea by the production of trypan red. Second, seeing that the chemical structure of atoxyl was such that it lent itself to modification in many ways, when professional chemists disagreed. Third, accepting Schaudinn's ideas that trypanosomes and spirochaetes were related and doggedly testing a large number of synthetic products, made in the search for trypanocidal drugs, in the treatment of spirochaetal diseases. For money Ehrlich was greatly indebted to Frau Franziska Speyer who built and endowed a new institute specifically for chemotherapeutic research, and it is doubtful if '606' could have been discovered without that money. The work involved was of a very expensive kind involving a team of experts and ancillary staff, the laborious synthesis and purification of complex organic molecules and tens of thousands of animal experiments. Luck, in Ehrlich's case, is an element difficult to evaluate but it seems to the writer that he was lucky to have available, when he started to prepare trypanocidal drugs, a substance that had undoubtedly a potent trypanocidal action – arsenic. It was a very good starting point, a better starting point than most of his successors in the field of chemotherapy have had. Another bit of luck has already been mentioned under the heading of 'intuitive genius' and yet surely it was great good fortune that, having developed a chemotherapeutic substance to cure experimental trypanosomiasis in rats, it was also effective, without modification, as a treatment of three important human diseases, relapsing fever, syphilis, and yaws.

From the introduction of salvarsan to the first reports of the successful treatment of streptococcal infections with the sulphonamides a quarter of a century elapsed. During that period there was considerable experimental work directed towards the chemotherapy of bacterial diseases but the successes achieved were minimal. None the less history cannot be devoted solely to the triumphs of mankind and it is therefore necessary that the work done in the chemotherapy of bacterial infection between 1910 and 1935 be briefly reviewed.

J. Morgenroth made what, at first, seemed a promising beginning to the chemotherapy of pneumonia. About 1911, starting from the widely held belief that quinine exerted a beneficial effect in pneumonia, he prepared a series of derivatives of hydrocupreine. By substituting various alkyl groups in place of the hydroxyl hydrogen of the parent substance he found that, as the length of the carbon chain increased, so the bactericidal activity also increased, up to a certain point. The pneumococcus and the diphtheria bacillus were found to be particularly sensitive to these hydrocupreine derivatives although the optimum length of the inserted carbon chain differed for each organism. Ethylhydrocupreine hydrochloride (optochin) seemed very promising when first tested in mice experimentally infected with a pneumococcus, which produced 100 per cent mortality in untreated controls. Optochin, if administered before the pneumococcus, prevented infection in about 90 per cent of animals and, if given after infection prevented the death of some 50 per cent of mice. However it was found that once infection in the mouse had become thoroughly established the toxic and therapeutic dose of optochin were too close to produce a useful effect. The drug none the less received several trials in man; among the first was a small trial by A. E. Wright in South Africans. But he, like other observers, found optochin of doubtful value and toxic effects such as blindness were soon reported quite frequently. Optochin never became established as a treatment of pneumonia in man, although it was used to irrigate infected wounds and as a local application to the conjunctiva for some time. It does however survive in the modern bacteriology laboratory as the basis of a most useful test for the identification of the pneumococcus. [228, 229]

Some progress was made in the 1920s in the chemotherapy of

leprosy, although it was hardly chemotherapy in Ehrlich's sense of the word. It was, in fact, the exploitation and improvement of a native remedy analagous to the use of quinine in malaria. It seems that the value of 'chaulmoogra oil', an oil derived from the seeds of trees growing in the tropical jungles of India and Burma, in the treatment of leprosy, had been appreciated from very ancient times. It was first introduced to western medicine in an article by F. J. Mouart, in 1854, entitled 'Notes on native remedies' in the first volume of the *Indian Annals of Medical Science*. Until the beginning of the twentieth century there was confusion as to just which jungle trees yielded the active oil but it was then settled that it came only from certain species of Hydnocarpus and Taraktogenos. Much work was done on the extraction and purification of the active principles, Chaulmoogric and hydnocarpic acids. Their ethyl esters seemed to be the most suitable derivatives for use in cases of leprosy. The drug was administered parentally. Since the leprosy bacillus cannot be cultivated *in vitro* and, at that time, it was not possible to infect experimental animals, the assessment of chaulmoogra preparations had to be made on human cases of leprosy. It was found by many workers, in different parts of the tropical world, that chaulmoogra derivatives were of considerable benefit in leprosy with upwards of 50 per cent of patients either cured or showing improvement.[230]

From about 1920 onwards some attempts were made at the chemotherapeutic treatment of tuberculosis. It had been known from the early studies of Robert Koch that some gold salts inhibited, even in high dilution, the growth of the tubercle bacillus *in vitro*, although without apparent effect in experimentally infected animals. None the less it was thought worthwhile to prepare a series of organic compounds containing gold for the treatment of tuberculosis but, in practice, they proved toxic and ineffective.

In 1924 H. Moellgaard, professor of physiology at the Royal Veterinary and Agricultural College, Copenhagen, introduced the double thiosulphate of gold and sodium, which had been known to chemists since 1845, under the name sanocrysin, into the treatment of tuberculosis. Moellgaard claimed to work on the principles of Ehrlich in chemotherapy and, although the quantity and quality of the experimental evidence he produced

was minimal, sanocrysin had extensive trial in the treatment of human tuberculosis and retained a diminishing place in the confidence of the medical profession right up to the introduction of streptomycin.[231] S. L. Cummins, David Davies professor of tuberculosis in Cardiff, reported some therapeutic activity in experimentally infected rabbits. Sanocrysin treated animals maintained their weight and survived longer than untreated controls but, when killed, showed definite tuberculous lesions. It also appeared that the effect of sanocrysin depended very much on the extent of the tuberculous infection and on the virulence and dose of the strain of bovine tubercle bacilli used to produce the infection.[232] In 1925 the British Medical Research Council published the preliminary results of a trial undertaken with sanocrysin supplied by Moellgaard. There were only twenty-two cases, treated at seven different medical schools, 'but it was the opinion of those observers who had most experience in dealing with consumption that the early cases of open tuberculous infection of the lungs did show some evident improvement though there was no dramatic benefit such as that seen with insulin or salvarsan in their corresponding diseases'. None the less sanocrysin was considered to be 'sufficiently encouraging to demand further clinical study'.[233] A further Medical Research Council report, published in 1926, based on 140 cases emphasized the dangers of sanocrysin therapy and could not point to any striking beneficial effect. But it nevertheless noted that 'in the opinion of some workers it has given indication that along the line of some such substance as sanocrysin there is definite hope of a drug treatment that will check the progress of a tuberculous infection and allow the patient's natural powers of resistance better play in finally arresting the disease'.[234] In fact recovery from tuberculosis, which was common on simple bed rest and general measures, depended upon so many variable factors that the accurate assessment of the value of any but the most dramatically effective chemotherapeutic agent was almost impossible.

In 1935 B. A. Peters and C. S. Short, from a tuberculosis sanatorium in Bristol, reported the results of their careful, but by no means ideal, trial of gold therapy in tuberculosis. They found that the differences between the control and treated groups of patients were 'statistically negligible' and concluded

that 'gold treatment is of no appreciable value'. They commented on their findings that 'the examination of our statistical results has been a painful shock, for we were convinced whilst carrying out this costly method of treatment that in chrysotherapy we had found a valuable aid; many of the cases seemed to do extremely well. But one tends to forget that many cases previously did extremely well even without gold. It is to us a sad reminder of the extreme fallibility of clinical judgement when exposed to the cruel light of a controlled statistical inquiry on a large number of cases'[235].

Probably the most useful chemotherapeutic substance for bacterial disease introduced between salvarsan and the sulphonamides were two 'urinary antiseptics' which, but for the plethora of antibiotics available would probably still be in use today. The substances alluded to, the discoveries of which were quite unconnected, are hexamine and mandelic acid. Hexamine was introduced into medicine at the end of the nineteenth century and, as L. P. Garrod wrote in 1935, it had 'been administered with the expectation of bactericidal action in almost every conceivable part of the body, including even the cerebro-spinal fluid and lungs'. There was a widespread failure to realize that hexamine itself has no bactericidal effect and it is only when it decomposes, which it does at an acid pH, that bactericidal fomaldehyde is liberated. When used as a urinary antiseptic, without controlling the pH of the urine, the results obtained were very erratic. But, as Garrod showed, if the pH of the urine was adjusted to 6.0, naturally excreted hexamine was powerfully bactericidal.[236]

The discovery of mandelic acid was due ultimately to the chance observation of H. F. Helmholz, of the Mayo clinic, that the urine of children on a ketogenic diet did not decompose on standing at room temperature. Attempts were therefore made to treat infections of the urinary tract by giving the patients a diet very rich in fat and poor in carbohydrate and considerable success achieved. A. T. Fuller, biochemist to the Bernhard Baron Memorial Research Laboratories at Queen Charlotte's Hospital, showed, by chemical analysis, that the bactericidal substance in the urine was b-hydroxybutyric acid. The administration of this substance by mouth was suggested as a treatment for urinary infections but it was quite ineffective because it was

metabolized in the body and not excreted in the urine. However, M. L. Rosenheim was encouraged to look for a keto or hydroxy acid which would be resistant to metabolism, non-toxic and bactericidal and found one in mandelic acid. Mandelic acid was only effective if the urine could be kept acid. At first it was prescribed with ammonium chloride and much work was done to ascertain the best preparation and the species of infecting bacteria which were sensitive to therapy. It was found that most of the common Gram negative bacilli which cause urinary tract infections, except species of Proteus, were amenable to treatment with mandelic acid and so also were infections due to Streptococcus faecalis.[237, 238] Rosenheim's claims for mandelic acid were soon fully confirmed and acclaimed as a very 'important therapeutic advance in the attack on a very common and disabling condition'. It is pleasant at this point, to meet once again H. C. Gram, now fifty-four years older than when he developed his immortal staining technique and in the last year of his life. In 1938 he published an account of a comparative trial of salol, hexamine and calcium mandelate in the treatment of pyelitis. There were twenty-seven patients in each of his groups and fourteen days' treatment with calcium mandelate sterilized the urine in twenty-three patients but hexamine did so in only eight.[240]

Soon after the introduction of salvarsan treatment for syphilis it was noticed, by workers in the Inoculation Department at St Mary's Hospital, that syphilitic lesions which were secondarily infected with pyogenic cocci healed just as well as those which were uninfected. This suggested to Sir Almroth Wright, who was studying the chemotherapeutic action of optochin about this time, that arsenicals might be active against pyogenic cocci as well as spirochaetes. His assistants, S. R. Douglas and L. Colebrook, therefore set to work to do some preliminary experiments.

Stewart Rankin Douglas was born in 1871 and studied medicine at St Bartholemew's Hospital. As a student he was not particularly distinguished, contented himself with the Conjoint qualification, in 1896, and entered the Indian Medical Service in 1898. At that time all I.M.S. officers attended a course at Netley and there Douglas met Almroth Wright. He must have made a favourable impression for, when Wright went to India

on a plague commission, he asked that Douglas be seconded to him. Douglas served in India and China but contracted amoebic dysentery and a liver abscess which forced his early retirement from the I.M.S. He immediately joined Wright, as his chief assistant, at St Mary's Hospital and worked with him until 1918. Douglas was then appointed director of the department of experimental pathology at the newly-formed Medical Research Council's laboratories at Hampstead. Here he headed, for the rest of his life, a team of some of the most distinguished medical microbiologists in Great Britain. It was Douglas, who was a keen sportsman with a wide knowledge of animals, who suggested distemper as a suitable virus disease for laboratory study and also the ferret as a laboratory animal. This work led directly to the proof of the viral aetiology of influenza. Douglas was elected F.R.S. in 1922. He died in 1936.[241]

Leonard Colebrook was born in 1882 and, as a student at St Mary's Hospital, came under the influence of Sir Almroth Wright. Soon after qualification, in 1906, he joined the staff of the Inoculation Department and, although becoming one of the staff of the Medical Research Council in 1920, continued to work at St Mary's until 1930. He worked on a variety of problems but his work during the First World War focused his interest on streptococcal infections. He was therefore well suited to become director of the Bernhard Baron Research Laboratories, at Queen Charlotte's maternity hospital, in 1930, since puerperal fever, a major obstetrical problem, was a predominantly streptococcal infection and such cases formed the largest single group of acute streptococcal infections. It was Colebrook who made the first clinical trials of the sulphonamides in this country. During the Second World War he undertook research on the prevention of infection and treatment of burns, at first in Glasgow and then at the Birmingham Accident Hospital, where he set up the Medical Research Council's Burns Unit. In 1945 he was elected F.R.S. and retired in 1948. During his retirement he wrote an excellent biography of his old master Almroth Wright. Colebrook died in 1967, aged 84.

Douglas and Colebrook found that both salvarsan and neosalvarsan in watery solution had a distinct bactericidal effect against staphylococci, at a dilution of 1 in 6,000, and that the serum of patients being treated with neosalvarsan, but not

salvarsan, acquired distinct bactericidal powers which lasted but a short time after the administration of the drug. They concluded, however, that the administration of neosalvarsan, or some similar preparation, might 'have a beneficial effect on wound or similar septicaemias, and might even strikingly assist in the sterilization of deep and irregular suppurating wounds'.[242] This possibility was to form Colebrook's main research interest for nearly twenty years.

These observations attracted the attention of C. S. Allison in America who repeated them using a haemolytic streptococcus instead of a staphylococcus and obtained similar results. Allison also went a step further and tested the effect of arsenicals on experimentally infected animals and achieved a degree of success; out of 21 control, untreated rabbits, 13 died, but of 25 treated rabbits only 8 died. In 1928 Colebrook published the further results of his own investigations which had in fact yielded but little. He confirmed that bactericidal properties were conferred on the serum of patients treated with arsenicals and worked out just how long this lasted after a given dose. He found that arsenicals, *in vitro*, exerted a toxic effect on leucocytes but, on the other hand, leucocytes taken from patients on treatment with arsenicals did not have their *in vitro* phagocytic powers materially decreased. Colebrook concluded that the arsenicals were worthy of a clinical trial in septicaemia since, in contrast to 'ordinary antiseptics' which had a higher affinity for leucocytes than bacteria, it ought to be possible, by adjustment of dosage, 'to maintain such a concentration of these arsenicals in a patient's blood as to avoid injury to the leucocytes while conferring a measure of bactericidal potency upon the blood fluids'.[243]

Soon after taking up his directorship of the Bernhard Baron Research Laboratories Colebrook and his colleague, Ronald Hare who later became professor of bacteriology at St Thomas's Hospital Medical School, began a clinical trial of the trivalent arsenicals in the treatment of puerperal fever. Cases were treated, not only at Queen Charlotte's Hospital, but at several other hospitals about London and all were demonstrated to be infected with a haemolytic streptococcus. In 1934 they reported the results of treatment in sixty-six cases. They divided their patients into three clinical groups of which one consisted of

H

twenty-eight cases, all with positive blood cultures but no clinical evidence of peritonitis. This group formed the best test for the efficacy of the arsenicals. Forty per cent of these patients survived and, at first, this seemed to point to a definite therapeutic success. However when they went through their old records of similar but untreated cases of puerperal fever, of the same clinical type, to make a retrospective control group, they were surprised to find an almost identical survival rate. Of their two other clinical groups, patients in whom infection appeared to be confined to the genital tract and those with generalized peritonitis, the former when treated with arsenicals, appeared to show a slight reduction in the frequency with which infection spread beyond the genital tract and the latter was quite unaffected by treatment.[244] These results must have been a great disappointment to Colebrook who had devoted so many years to work along these lines, particularly when the initial *in vitro* work had seemed promising. The results he and Hare obtained showed clearly that arsenicals were useless in the treatment of human streptococcal infection. Happily the introduction of the sulphonamides the very next year virtually solved the problem of acute haemolytic streptococcus infection in man.

The synthetic dye industry formed the basis of much chemotherapeutic research and it was from this source that the first major advance after Ehrlich's '606' came with the discovery of the sulphonamides. The rationale behind this approach was, firstly, that in Ehrlich's hands synthetic dyes had been shown to be possible sources of chemotherapeutic agents, for example, trypan red. Secondly that the chemists of the dye industry were constantly producing new dyes, in the search for dyes with advantages as dyes over existing compounds, and, therefore, a huge range of new chemicals was readily available for testing for antibacterial activity. The I. G. Farbenindustrie, at Wuppertal-Elberfeld, set up a department for just this purpose and, in 1927, Gerhard Domagk who was later to discover the chemotherapeutic activity of the sulphonamides, was appointed director of research in experimental pathology and bacteriology.

Many dyes were soon found to be capable of inhibiting the growth of bacteria at quite high dilutions *in vitro* and some compounds, particularly among the acridine dyes, such as

acriflavine and proflavine, received extensive trials as local applications to infected wounds. They were also tried as true chemotherapeutic agents both in experimentally infected animals and in man without producing results of practical value.[245, 246]

The discovery of the sulphonamides was a direct result of the I. G. Farbenindustrie policy of routinely screening dyes and other compounds produced by their chemists for antibacterial activity. All newly synthesized substances were tested for activity against a range of bacterial species *in vitro* and tested for toxicity in animals. Any compound which seemed promising, as a result of these tests, was tried for chemotherapeutic action in experimentally infected animals. The azo-dyes had been found to have antibacterial activity as far back as 1913 and some antiseptics and a few urinary antiseptics, of little value in practice, had been developed. Azo-dyes containing sulphonamide groups had been synthesized in 1910 by H. Horlein and his colleagues and some time, much later, Domagk noted that sulphonamide containing compounds had some activity against streptococcal infection in mice. The circumstances and exact date of this initial observation are not clear. However, around 1930, Domagk's chemical colleagues were busy producing many sulphonamide-containing azo-compounds for him to test for antistreptococcal activity in mice. At any rate the first German patent on a new dye, 4-sulphamido-2-4-diaminoazobenzol (Prontosil), was dated 25 December 1932 and its activity in mice infected with streptococci was known at that time. Nor was this the only active compound discovered. Horlein, in a paper read to the Royal Society of Medicine, in London in October 1935, claimed that 'many other azo-compounds substituted in a certain way with sulphonamide groups, displayed an almost selective chemotherapeutic action in the streptococcal sepsis of mice'.[247] The I. G. Farbenindustrie workers however were either ignorant of, or did not choose to report, that the simple, unpatentable sulphanilamide, the sulphonamide group on to which they fitted various chemical groups in producing Prontosil, etc., was just as active as any of their more complicated molecules. Moreover although aware of the antistreptococcal activity of Prontosil, by the end of 1932, this information was not immediately published; it was not until two years later, in

February 1935, that Domagk published the first account of his experimental work.

Domagk's original report is a brief paper in the Deutsche Medicinische Wochenschrift. After reviewing the earlier work on the chemotherapy of bacterial infections he reported in detail a single mouse experiment. He inoculated mice intraperitoneally with a broth culture of a haemolytic streptococcus and then treated twelve of them with different doses of Prontosil directly into the stomach. All his control, untreated mice died within three days but all twelve treated animals were alive and well on the seventh day; whether they survived indefinitely Domagk does not record. He also claimed that infected rabbits with joint abscesses and endocarditis were improved by Prontosil but had not made a proper controlled trial.[248]

In the same issue of the journal are two clinical reports of the use of Prontosil in man. The first, from P. H. Klee and H. Romer, working in a hospital in Wuppertal-Elberfeld, described how they had tried prontosil in a considerable variety of streptococcal infections including sore throats, lymphadenitis, erysipelas, endocarditis and polyarthritis. Few bacteriological details were given and the paper merely records a general favourable impression of the action of Prontosil. It can hardly be said to contribute much towards establishing the usefulness of the drug in man. The other paper, by H. T. Schreus, from a skin clinic in Dusseldorf, reported his impression of the value of Prontosil in the treatment of erysipelas, which was a good choice since it is a well-defined purely streptococcal disease. No statistics were given but it was stated that, generally, within forty-eight hours on Prontosil, the patients temperature was normal and the development of the skin rash had been brought to a standstill.

In fact Prontosil, or rather a more soluble derivative known as 'Streptozon', had been tried in at least one case of human infection almost two years before Domagk's first report of his work in mice. Foerster reported a case, at a dermatological meeting in Dusseldorf, in May 1933, of a ten-month-old boy who was almost dead from staphylococcal pyaemia, with a positive blood culture, whose temperature fell to normal within four days and who then recovered.[249]

Any hopes that the firm of I. G. Farbenindustrie might have

had, of having a monopoly of an important chemotherapeutic agent, were soon dashed to the ground. The Trefouels and their colleagues, at the Pasteur Institute, began work on the Prontosil molecule substituting other chemical groups into different parts of it. It soon became apparent to them that derivatives in which the sulphanilamide part of the molecule was left intact were all active against streptococci but, if the sulphanilamide part of the molecule was removed, the derivative was inactive. They naturally tried the effect of the simple substance p-amino-phenylsulphonamide and found that it was just as active as Prontosil in the treatment of streptococcal infection in mice. This discovery they announced on 23 November 1935, nine months after Domagk's original communication.[250] In March 1935 Sir Henry Dale, director of the National Institute for Medical Research, in England, asked for a trial supply of Prontosil and this was entrusted to Colebrook to test. His first result, of an experiment in mice, was disappointing indeed – some of his treated mice lived rather longer than the controls but there were practically no survivors. Colebrook regarded his results as 'negative, failing to confirm Domagk's claim'. At this point Colebrook heard that Dr G. A. H. Buttle, of the Wellcome laboratories, had had rather more success but he, instead of using freshly isolated strains of streptococci, was using a strain which had been rendered highly mouse-pathogenic by passage through a series of twenty-three mice. Colebrook immediately repeated his experiments using a specially mouse-passaged strain and at once began to get striking curative results. Colebrook then went on to make a careful study of the action of Prontosil in infected mice, showed that the drug exerted a bacteristatic rather than bactericidal action and that it was not toxic to leucocytes. He also confirmed that sulphanilamide was just as effective as Prontosil.

But the most useful part of his work was his clinical assessment in the human patient. For such work Colebrook was well fitted having at his command abundant cases of puerperal fever as well as previous experience of testing arsenicals in this condition. There had, it is true, been a number of German clinical reports on the value of Prontosil but, Colebrook pointed out, 'their evidential value must be regarded as small since, in most cases, the recovery of patients was unhesitatingly ascribed to

the treatment, and too little allowance is made for the tendency to spontaneous cure of these infections. The bacteriological and clinical data supplied are nearly always very scanty, e.g. we are not told whether the cases were all infected by haemolytic streptococci. . . .' The results of his mouse experiments had made Colebrook very doubtful that Prontosil would have any curative action in puerperal fever cases but, since all the German clinical reports had been favourable and they had at least demonstrated that the drug was safe, he agreed to try Prontosil on a group of puerperal fever cases, all proved to be infected with a group A streptococcus (except two). His first clinical trial involved thirty-eight cases but he was extremely cautious in the interpretation of his results. Of these patients eighteen were excluded from the trial proper because it was felt, on clinical grounds, that they would probably have got better anyway without treatment. But there was a group of sixteen patients which left him with 'the impression that the drug had in all probability hastened or determined recovery from the infection'. This judgement was made on the basis of his previous experience, the severity of the illness and the rapid improvement that followed a few doses of Prontosil. There were however five patients in which Prontosil was unhelpful and three of them died. None the less out of thirty-eight patients in all, only three had died, a mortality of 8 per cent. Among the thirty-eight cases of puerperal fever treated immediately prior to the Prontosil trial ten had died, a mortality of 26 per cent. And, during the previous four years, the mortality had averaged 22 per cent and had never been lower than 18 per cent at any time. But Colebrook also noted that Prontosil-treated patients continued to excrete haemolytic streptococci in the lochia for about as long as untreated patients who recovered.

The German publications had emphasized the safety of Prontosil and certainly Colebrook found no serious complications. He did draw attention to a definite toxic effect on the kidneys – a significant number of his patients were found to have red blood cells, albumen and casts in the urine. He also recorded three cases of a more interesting complication, the development of sulphaemoglobinaemia in three patients.[251]

During the next few years Prontosil or sulphanilamide were to be found valuable in the treatment of all the common

infections of man with a group A streptococcus, although accurate assessment in the case of the milder types of infection, which ran more variable courses, was more difficult than with severe infections such as puerperal fever. Infections with other groups of streptococci were, on the whole, more resistant. Sulphanilamide was soon tried in a wide variety of other bacterial infections, both experimental and naturally occurring. Buttle showed, in 1936, that sulphanilamide protected mice against infection with the meningococcus and the drug was tried in cases of meningitis in man. The early series of cases was small and there was a reluctance not to administer meningococcus antiserum as well, thus making assessment more difficult. However F. F. Schwentker and his colleagues, in America, soon showed that sulphonamide-treated patients had about half the mortality of serum-treated patients.[252] Linser, in Germany, in 1936, showed that sulphonamides were valuable in the treatment of gonorrhoea and a number of workers demonstrated dramatic curative effects in pyelitis due to *Escherichia coli*. Staphylococcal infections were found to be to some extent amenable to treatment with sulphonamides, although not as satisfactorily as streptococcal infection.[253]

One of the most widespread and fatal of all acute infectious diseases in the 1930s was pneumococcal pneumonia. Sir William Osler had called it the 'captain of the men of death'. The relatively close relationship between the streptococci and the pneumococcus gave grounds for hope that the sulphonamides might be of value against pneumococcal infection. Domagk, in 1935, found that Prontosil had some slight action against pneumococcus type III but none against types I and II and, during the two following years, it became clear that Prontosil and sulphanilamide, although they might show some activity against some types of pneumococci, were not of much clinical use. G. A. H. Buttle and his colleagues, at the Wellcome laboratories, were among the most active workers on sulphonamides in the months immediately following their discovery. They were very busy, not only testing sulphanilamide against a wide range of bacterial infections in experimental animals, but also in synthesizing and testing new related compounds. In June 1937 they reported preliminary results with two new compounds 4:4 diaminodiphenyl sulphone and 4:4 dinitro-

diphenylsulphone. The former was one hundred times more active than sulphanilamide against streptococci although, at the same time, twenty-five times as toxic. But preliminary tests showed that 4:4 diaminodiphenylsulphone was much more effective in prolonging the life of mice infected with pneumococci than sulphanilamide although, at that time it was uncertain if mice could be completely cured.[254]

Another research team active in the field of chemotherapy against the pneumococcus was the firm of May and Baker Ltd. led by A. J. Ewins who had enlisted the help of Lionel E. H. Whitby, of the Middlesex Hospital, to test their synthetic products against various infections in mice. One of these products, 2-p-aminobenzenesulphonamidopyridine, designated in Whitby's original paper and for ever afterwards, for short, as 'M and B 693' afforded striking protection to mice infected with a variety of pneumococcal types. Batches of mice inoculated with 50,000 pneumococci normally died within less than twenty-four hours but, when treated with M and B 693, survived from four to seven days.[255] A few weeks later G. M. Evans and W. F. Gaisford, physicians to Dudley Road Hospital, Birmingham, published the dramatic results they had obtained in treating acute lobar pneumonia in man with M and B 693. Between March and June 1938 over 200 cases of lobar pneumonia were admitted to Dudley Road Hospital and half of these (those patients admitted under the care of Evans and Gaisford) were treated with M and B 693 and the other half (admitted under the care of other physicians) were treated in the usual way, without M and B 693. The case mortality in the control group was 27 per cent but in the M and B 693 treated only 8 per cent – the days of pneumonia as the 'captain of the men of death' were over.[256]

The discovery of M and B 693 was undoubtedly the greatest advance in chemotherapy since the introduction of the sulphonamides and yet the name of the man who could claim to have been most responsible for its production does not appear on a single paper on the subject, except as a formal acknowledgement at the foot of Whitby's paper. This man was A. J. Ewins, director of research at May and Baker Ltd. He was, as might be imagined, 'a man of retiring disposition, modest and simple in his tastes and habits, straightforward in all his

dealings, and in his ways of thinking'. He had worked his way up from the humble post of technical assistant at the Wellcome laboratories, to which he went straight from school at the age of 17, in 1899. He was active in research at the Wellcome laboratories until 1916 when he joined May and Baker Ltd. Thereafter his personal publications almost ceased and his name was rarely given more than a modest place in the acknowledgements at the end of his colleagues papers. Happily it was well understood, by those working in the field of chemotherapy, that his contributions were of the first importance and he was elected F.R.S. in 1943. He died in 1958.[257]

References

1 Bonomo, Dr, translated in *The Medical Works of Richard Mead* (Edinburgh, 1765). Vol. I, p. 241.
2 Dobell, C. (1932). *Antony van Lecuwenhoek and His Little Animals*, London.
3 Hoepple, R. (1959). *Parasites and Parasitic Infections in Early Medicine and Science* (Singapore), p. 86.
4 Rolleston, J. D. (1934). *Pro. Roy. Soc. Med., 18*, 1.
5 Bulloch, W. (1938). *The History of Bacteriology* (London), p. 159.
6 Ackernecht, E. H. (1964). *J. Hist. Med., 19*, 131.
7 Henle, J. (1938). Translated in *Bull. Hist. Med., 6*, 1.
8 Sanderson, J. B. (1875). *Trans. Path. Soc. Lond., 26*, 255.
9 Goodsir, J. (1842). *Edin. Med. Surg. J.*
10 Hassall, A. H. (1853). *The Lancet, 1*, 338.
11 Liebig, J. (1846). *The Lancet, 2*, 420.
12 Simon, J. (1850). *The Lancet, 2*, 230.
13 Wilsher, Dr (1849). *The Lancet, 1*, 98.
14 Bigelow, J. (1867). *Modern Inquiries* (Boston), p. 297.
15 Villemin, J. A. (1868). Translated in Major's *Classic Descriptions of Disease*, 3rd edition (Oxford, 1948), p. 66.
16 Marcet, W. (1867). *Medico-Chirurgical Trans., 50*, 1.
17 Foster, W. D. (1965). *A History of Parasitology* (Edinburgh), p. 74.
18 Lister, J. (1909). *Collected Papers*, Vol. II, p. 2.
19 Wells, T. S. (1864). *Medical Times and Gazette*.
20 Beale, L. S. (1872). *Disease Germs, Their Nature and Origin* 2nd edition, London.
21 Bloomfield, A. L. (1958). *A Bibliography of Internal Medicine* (Chicago), p. 233.
22 Davaine, C. J. (1889). *L'oeuvre* (Paris), p. 31.
23 Koch, R. (1876). Translated in Brock's *Milestones in Microbiology* (London 1961), p. 90.
24 Pasteur, L. (1933). *L'oeuvres*, Vol. 6, p. 112.
25 Pasteur, L. (1933). Ibid., p. 161.
26 Pasteur, L. (1933). Ibid., p. 151.

27 Pasteur, L. (1933). Ibid., p. 147.
28 Pasteur, L. (1933). Ibid., p. 291.
29 Duclaux, E. (1920). *Pasteur, the History of a Mind*, translated by E. F. Smith and F. Hedges (Philadelphia), p. 275.
30 Pasteur, L. (1933). *L'oeuvre*, Vol. 6, p. 346.
31 Pasteur, L. (1933). Ibid., p. 370.
32 Galtier, V. (1881). *Bull. Acad. Med., 10*, 90.
33 Pasteur, L. (1933). *L'oeuvres*, Vol. 6, p. 559.
34 Pasteur, L. (1933). Ibid., p. 603.
35 Metchnikoff, E. (1959). *Souvenirs* (Moscow), p. 286.
36 Clyde, D. G. (1962). *History of the Medical Services in Tanganyika* (Dar-es-Salaam), p. 20.
37 Cook, A. R. (1945). *Uganda Memories* (Kampala), p. 100.
38 Letter from Hodges, A. D. P. to Bruce, D., 9 August 1907 (in Makerere Medical Library, Kampala).
39 Koch, R. (1960). Translated in Doetsch's *Microbiology – Historical Contributions* (New Brunswick), p. 67.
40 Koch, R. (1880). *Investigations into the Etiology of Traumatic Infective Diseases*, translated by W. W. Cheyn, London.
41 Koch, R. (1884). *Bacteria in Relation to Disease*, translated by W. W. Cheyn (London), p. 32.
42 Hitchens, A. P. and Leikind, M. C. (1939). *J. Bact., 37*, 485.
43 Bulloch, W. (1938). *The History of Bacteriology* (London), p. 230.
44 Koch, R. (1884). *Bacteria in Relation to Disease*, translated by W. W. Cheyne (London), p. 493.
45 Anno. (1881). *The Lancet, 1*, 547.
46 Koch, R. (1882). Translated in Brock's *Milestones in Microbiology* (London), p. 109.
47 Foster, W. D. (1961). *A Short History of Clinical Pathology* (Edinburgh), p. 60.
48 Scott, H. H. (1939). *A History of Tropical Medicine* (London), Vol. II, p. 666.
49 Pasteur, L. (1933). *L'oeuvres*, Vol. 6, p. 541.
50 Koch, R. (1881). Translated in W. W. Cheyne's *Microparasites in Disease* (London, 1886), p. 1.
51 Lewis, T. R. (1888). *Physiological and Pathological Researches* (London), p. 1.
52 *Brit. Med. J.* (1886), *2*, 1294.
53 Lewis, T. R. (1888). *Physiological and Pathological Researches* (London), p. 329.
54 Finkler, D. and Prior, J. (1884). *Dtsch. Med. Wschr. 10*, 579.
55 Lewis, G. W. (1885). *Buffalo Med. and Surg. J.*
56 Koch, R. (1891). *Dtsch. Med. Wschr., 17*, 101.
57 *Brit. Med. J.*, (1890), *2*, 1198.

58 Koch, R. (1890). *Brit. Med. J.*, *2*, 380.

59 Koch, R. (1890). Ibid., *2*, 1193.

60 *Brit. Med. J.* (1890), 2, 1197.

61 Ibid., *2*, 1200.

62 Koch, R. (1902). *Trans. Brit. Congr. Tuberculosis*, *3*, 93.

63 Koch, R. (1902). Ibid, *1*, 25.

64 Cushing, H. (1926). *The Life of Sir William Osler* (Oxford), Vol. I., p. 114.

65 Billroth, T. (1877). *Lectures on Surgical Pathology and Therapeutics* (London), Vol. 1, p. 136.

66 Fehleisen, F. (1886) in W. W. Cheyne's *Bacteria in Relation to Disease* (London), p. 261.

67 Ogston, W. H., Cowen, H. H. and Smith, H. E. (1943). *Alexander Ogston, K.C.V.O.* (Aberdeen), p. 98.

68 Ogston, A. (1881). *Brit. Med. J.*, *1*, 369.

69 Lister, J. (1881). *The Lancet*, *2*, 695.

70 Ogston, A. (1882). *Jour. Anat.*, *16* and *17*.

71 Rosenbach, J. (1884) in W. W. Cheyne's *Bacteria in Relation to Disease* (London), p. 397.

72 Bloomfield, A. C. (1958). *A Bibliography of Internal Medicine – Communicable Diseases* (Chicago), p. 107.

73 Bulloch, W. (1923) in M. R. C. *Diphtheria, its Bacteriology, Pathology and Immunology* (London), p. 38.

74 Loeffler, F. (1913) in G. H. F. Nuttall and G. S. Graham Smith's *The Bacteriology of Diphtheria* (Cambridge), p. 37.

75 Magnin, A. (1885). *Bacteria* translated by G. M. Sternberg, 2nd edition, p. 400.

76 Gaffky, G. (1884) in W. W. Cheyne's *Microparasites and Disease* (London), p. 205.

77 Foster, W. D. (1961). *A Short History of Clinical Pathology* (Edinburgh), p. 70.

78 Bloomfield, A. L. (1958). *A Bibliography of Internal Medicine – Communicable Diseases* (Chicago), p. 89.

79 Sternberg, G. M. (1896). *A Textbook of Bacteriology* (New York), p. 300.

80 White, B. (1938). *The Biology of the Pneumococcus* (New York), p. 3.

81 Fraenkel, C. (1891). *Textbook of Bacteriology*, translated by J. H. Linsley, p. 306.

82 Cheyne, W. W. (1880). *Brit. Med. J.*, *2*, 124.

83 Bloomfield, A. L. (1958). *A Bibliography of Internal Medicine – Communicable Diseases* (Chicago), p. 185.

84 *The Lancet* (1879), 2, 666.

85 Bloomfield, A. L. (1958). A Bibliography of Internal Medicine – Communicable Diseases (Chicago), p. 172.
86 Councilman, W. T., Mallory, F. B. and Wright, J. H. (1898). Epidemic Cerebrospinal Meningitis and its Relation to other forms of Meningitis (Boston).
87 Bruce, D. (1887). Practitioner, 39, 161.
88 Bruce, D. (1888). Ibid., 40, 241.
89 Escherich, T. (1886). Die Darmbakterium des Sauglings, Stuttgart.
90 Metchnikoff, O. (1921). Life of Elie Metchnikoff, London.
91 Ewart, J. C. (1880). Brit. Med. J., 4, 442.
92 Shafer, A. E. (1882). Ibid., 2, 573.
93 Metchnikoff, E. (1884). Translated in Brock's Milestones in Microbiology (London, 1961), p. 132.
94 Metchnikoff, E. (1893). Lectures on the Comparative Pathology of Inflammation, London.
95 Ziegler, E. (1883). A Textbook of General Pathological Anatomy and Pathogenesis, New York.
96 Bulloch, W. (1938). The History of Bacteriology (London), p. 256.
97 Koch, R. (1890). Brit. Med. J., 2, 378.
98 Lister, J. (1890). Ibid., p. 380.
99 Ruffer, A. (1890). The Lancet, 2, 564.
100 Loeffler, F. (1913) in Nuttall and Graham-Smith's The Bacteriology of Diphtheria (Cambridge), p. 48.
101 Behring, E. and Kitasato, S. (1890). Brit. Med. J., 2, 1395.
102 Behring, E. (1891). Ibid., 2, 406.
103 Nuttall, G. H. F. (1924). Parasitology, 16, 229.
104 Roux, E. (1891). Brit. Med. J., 2, 378.
105 Fildes, P. (1956). Biog. Memoirs F.R.S., 2, 237.
106 Bordet, J. (1895). Ann. Inst. Pasteur, 9, 462.
107 Oakley, C. L. (1962). Biog. Memoirs F.R.S., 8, 19.
108 Durham, H. E., (1896). Proc. Roy. Soc., 54, 224.
109 Foster, W. D. (1961). A Short History of Clinical Pathology (Edinburgh), p. 75.
110 Bordet, J. (1898). Ann. Inst. Pasteur, 12, 688.
111 Bulloch, W. (1938). The History of Bacteriology (London), p. 363.
112 Muir, R. (1916). Jour. Path. Bact., 20, 350.
113 Ehrlich, P. (1956) in The Collected Papers of Paul Ehrlich (London), Vol. I, p. 311.
114 The Lancet (1896). 2, 182.
115 Marquardt, M. (1949). Paul Ehrlich, London.
116 Ehrlich, P. (1957) in The Collected Papers of Paul Ehrlich (London), Vol. II.
117 Bordet, J. (1909) in Studies in Immunity, translated by F. P. Gay, New York.

118 Arrhenius, S. (1907). *Immunochemistry*, New York.
119 *Clin. Med. Surg.* (1933), *40*, 67.
120 *Brit. Med. J.* (1885), *1*, 824, 1114, 1122.
121 Ibid. (1885), *2*, 33, 165, 806.
122 Waksman, S. A. (1964) *The Brilliant and Tragic Life of W. M. W. Haffkine*, New Brunswick.
123 Haffkine, W. M. W. (1892). *Comp. Rend. Soc. Biol.*, *4*, 635, 671, 740.
124 Haffkine, W. M. W. (1895). *Brit. Med. J.*, *2*, 1541.
125 Haffkine, W. M. W. (1897). Ibid., *1*, 1461.
126 Wright, A. E. (1891). Ibid., *2*, 641.
127 Wright, A. E. (1893). Ibid., *1*, 227.
128 Wright, A. E. (1896). *The Lancet*, 2, 807.
129 Wright, A. E. (1897). *Brit. Med. J.*, *1*, 256.
130 Wright, A. E. (1909). *Studies on Immunization* (London), p. 26.
131 Wright, A. E. (1901). *The Lancet*, *2*, 715.
132 Wright, A. E. (1904). *A Treatise on Antityphoid Inoculation* (London), p. 43.
133 Topley, W. W. C. (1933). *An Outline of Immunity* (London), p. 9.
134 Wright, A. E. (1902). *The Lancet*, *1*, 964.
135 Oliver, W. W. (1941). *The Man who Lived for Tomorrow*, New York.
136 Park, W. H. (1914). *Arch. Ped.*, *34*, 481.
137 Park, W. H. (1914). *J.A.M.A.*, *63*, 859.
138 Park, W. H. (1922). Ibid., *79*, 1584.
139 *The Lancet* (1924), *1*, 139.
140 Ibid. (1924), *2*, 971, 1086.
141 Ramon, G. (1923). *Bull. Acad. Roy. Med. Belg.*, *3*, 740.
142 Glenny, A. T. (1921). *J. Hyg. Camb.*, *20*, 176.
143 Glenny, A. T. and Hopkins, B. (1923). *Brit. J. Exp. Path.*, *4*, 283.
144 Ramon, G. (1923). *C. R. Acad. Sci.*, *177*, 1338.
145 Ramon, G. (1924). *Ann. Inst. Pasteur*, *38*, 1.
146 Ramon, G. (1950). *Le Principe des Anatoxines et ses Applications* (Paris), p. 74.
147 Kervan, R. 1962). *Albert Calmette et le B.C.G.*, Paris.
148 *The Lancet* (1961), *1*.
149 Calmette, A. (1923). *Tubercle Bacillus Infection and Tuberculosis in Man and Animals* (Baltimore), p. 638.
150 L. A. (1928). *Brit. Med. J.*, *1*, 793.
151 L. A. (1930). Ibid., *2*, 694.
152 Wilson, G. S. (1967). *The Hazards of Immunization* (London), p. 66.

153 Richet, C. (1913). Anaphylaxis, translated by J. M. Bligh, Liverpool.
154 Pirquet, C. F. and Schick, B. (1951). Serum Sickness, translated by B. Schick, Baltimore.
155 Pirquet, C. F. (1906). Munch. Med. Wschr., 30, 1457. Translated in Gell and Coombs Clinical Aspects of Immunology, Oxford, 1963.
156 Besredka, A. (1919). Anaphylaxis and Anti-anaphylaxis and their Experimental Foundations, translated by S. R. Gloyne, St Louis.
157 Schultz, F. (1910). J. Pharmacol and Exp. Therapeutic, 1, 549.
158 Dale, H. (1913). Ibid., 44.
159 Noon, L. (1911). The Lancet, 1, 1572.
160 Freeman, J. (1911). Ibid., 2, 814.
161 Zinsser, H. (1931). Resistance to Infective Diseases, 4th edition (New York), p. 443.
162 Prausnitz, C. and Kustner, H. (1921). Centralbl. F. Bakt., 86, 160. Translated in Gell and Coombs Clinical Aspects of Immunology, Oxford, 1963.
163 Klemperer, G. and Klemperer, F. (1891). Berlin Kln. Wschr., 28, 833.
164 Washbourn, J. W. (1896). Jour. Path. Bact., 3, 214.
165 Washbourn, J. W. (1897). Brit. Med. J., 1, 510.
166 Washbourn, J. W. (1899). The Lancet, 1, 954.
167 Washbourn, J. W. and Eyre, J. W. H. (1899). Brit. Med. J., 2, 1248.
168 Mennes, F. (1898). Zeitschr. f. Hyg. v. Inf., 26, 413.
169 Anders, J. M. (1904). J.A.M.A., 43, 1777.
170 Neufeld, F. (1902). Zeitschr. f. Hyg. v. Inf., 40, 54.
171 Neufeld, F. (1910).
172 Averoy, O. T. (1932). Ann. Int. Med., 6, 1.
173 The Lancet (1941), 1, 588.
174 Hayes, W. (1966). J. Hyg. Camb., 64, 178.
175 Griffith, F. (1928). Ibid., 27, 113.
176 Avery, O. T. (1932). Ann. Int. Med., 6, 1.
177 Hammock, A. (1895). Ann. Inst. Pasteur, 9, 593.
178 Moser, P. (1902). Wein. klin. Wschr., 15, 1053.
179 Weaver, G. H. (1904). J. Inf. Dis., 1, 91.
180 Brown, J. H. (1919). The Use of Blood Agar for the Study of Streptococci, New York.
181 Jochmann, G. (1905). Ztschr. f. Klin. Med., 56, 316.
182 Dick, F. and Dick, G. (1924). J.A.M.A., 82, 301.
183 Dochez, A. R. (1924). Ibid., 82, 542.
184 Schultz, W. and Carlton, W. (1918). Ztschr. f. Kinderkr., 17, 328.
185 Griffith, F. (1926). J. Hyg. Camb., 25, 385.

186 Lancefield, R. (1933). *J. Exper. Med.*, *57*, 571.
187 Griffith, F. (1934). *J. Hyg. Camb.*, *34*, 542.
188 Fowler, J. K. (1880). *The Lancet*, *2*, 933.
189 Coburn, A. F. (1932). *J. Exper. Med.*, *56*, 609.
190 Schlesinger, B. and Signy, A. G. (1933). *Quart. J. Med.*, *22*, 255.
191 Todd, E. W. (1932). *Brit. J. Exper. Path.*, *13*, 248.
192 Myers, W. K. and Keeper, C. S. (1934). *J. Clin. Invest.*, *13*, 155.
193 Durham, H. E. (1898). *Brit. Med. J.*, *2*, 1797.
194 Durham, H. E. (1898). *The Lancet*, *1*, 154.
195 Durham, H. E. (1900). *J. Exper. Med.*, *5*, 353.
196 Castellani, A. (1960). *Men, Microbes and Monarchs*, London.
197 Foster, W. D. (1961). *A Short History of Clinical Pathology* (Edinburgh), p. 75.
198 Hughes, M. L. (1897). *Mediterranean, Malta or Undulant Fever* (London), p. 52.
199 Horrocks, W. H. (1905). *Pro. Roy. Soc. B.*, *76*, 378.
200 Bang, B. (1897). Reprinted in *Bernard Bang: Selected Works*, (London, 1936).
201 Smith, T. and Fabyan, M. (1912). *Centralbl. f. Bakt.*, *61*, 549.
202 Larson, W. P. and Sedgwick, J. P. (1913). *Amer. J. Dis. Child.*, *6*, 326.
203 Evans, A. (1918). *J. Inf. Dis.*, *22*, 580.
204 Bevan, L. E. W. (1925). *Brit. Med. J.*, *1*, 554.
205 Keefer, C. S. (1924). *Bull. Johns Hop. Hosp.*, *35*, 6.
206 *Brit. Med. J.* (1910), *2*, 1501.
207 Colebrook, D. (1935) M. R. C., special report number 205.
208 Twort, F. W. (1915). *The Lancet*, *2*, 1241.
209 Nature (1949). *163*, 984.
210 d'Herelle, F. (1917). *Comp. Rend. Adad. Sci.*, *165*, 373.
211 d'Herelle, F. (1911). Ibid., *152*, 1413.
212 d'Herelle, F. (1914). *Ann. Inst. Pasteur*, *28*, 280.
213 d'Herelle, F. (1926). *The Bacteriophage and its Behaviour*, Baltimore.
214 Felix, A. and Pitt, R. M. (1934). *Jour. Path. Bact.*, *38*, 409.
215 Felix, A. and Pitt, R. M. (1934). *The Lancet*, *2*, 189.
216 Craigie, J. and Yen, C. H. (1938). *Canad. J. Publ. Hlth.*, *29*, 448 and 484.
217 Ehrlich, P. (1957). *The Collected Papers of Paul Ehrlich* (London, 1957), Vol. I.
218 Ehrlich, P. (1959). Ibid., Vol. III.
219 Ehrlich, P. and Hata, S. (1911). *The Experimental Chemotherapy of Spirillosis* (London), p. 120.

220 Laveran, A. and Mesnil, F. (1907). *Trypanosomes and Trypano-somiasis* (London), p. 282.
221 Thomas, H. W. (1904). *Brit. Med. J., 1,* 1337.
222 Thomas, H. W. (1905). Ibid., *1,* 1140.
223 Ehrlich, P. and Hata, S. (1911). *The Experimental Chemotherapy of Spirillosis* (London), p. 134.
224 Marquardt, M. (1949). *Paul Ehrlich,* p. 163.
225 Ferryrolles, F. (1908). *The Lancet, 1,* 1575.
226 Ehrlich, P. and Hata, S. (1911). *The Experimental Chemotherapy of Spirillosis* (London), p. 91.
227 L. A. (1910). *The Lancet, 2,* 740.
228 Findlay, G. M. (1930). *Recent Advances in Chemotherapy,* 1st edition (London), p. 493.
229 Wright, A. E. (1914). *On Pharmacotherapy and Preventive Inoculation Applied to Pneumonia in the African Native.*
230 Findlay, G. M. (1930). *Recent Advances in Chemotherapy,* 1st edition (London), p. 430.
231 Moellgaard, H. (1925). *Brit. Med. J., 1,* 643
232 Cummins, S. C. (1926). *Brit. J. Exper. Path.,* 7, 47.
233 *Brit. Med. J.* (1925), *2,* 158.
234 Ibid. (1926), *2,* 158.
235 Peters, B. A. and Short, C. S. (1935), *The Lancet, 2,* 11.
236 Garrod, L. P. (1935). *St Barts. Hosp. Rep., 68,* 84.
237 Findlay, G. M. (1939). *Recent Advances in Chemotherapy,* 2nd edition (London), p. 474.
238 Garrod, L. P. (1959). *Chemotherapy for Infections of the Urinary Tract* (Edinburgh), p. 17.
239 *Brit. Med. J.* (1935), *2,* 1096.
240 Gram, H. C. (1938). *Acta. Med. Scand., 94,* 615.
241 Obituary notice (1936). *Jour. Path. Bact., 42,* 515.
242 Douglas, S. R. and Colebrook, L. (1916). *The Lancet 1,* 181.
243 Colebrook, L. (1928). M.R.C. special report series No. 119.
244 Colebrook, L. and Hare, R. (1934). *The Lancet, 1,* 388.
245 Findlay, G. M. (1930). *Recent Advances in Chemotherapy,* 1st edition (London), p. 482.
246 Churchman, J. W. (1925), *85,* 1849.
247 Horlein, H. (1935). *Pro. Roy. Soc. Med., 29,* 321.
248 Domagk, G. (1935). *Deutsche Med. Wschr., 61,* 250.
249 Foerster, Dr (1933). *Zbl. Haut v. Geschlekr., 45,* 549.
250 Trefouel, J. Nitti, F., Bovet, D. (1935). *C. R. Soc. Biol., 120,* 756.
251 Colebrook, L. (1936). *The Lancet, 1,* 1279.
252 Schwentker, F. F. (1937). *J.A.M.A., 108,* 1407.
253 Findlay, G. M. (1939). *Recent Advances in Chemotherapy,* 2nd edition (London), p. 398.

254 Buttle G. A. H., Stephenson, D., Smith, S., Dewing, T. and Foster, G. E. (1937). *The Lancet*, *1*, 1331.
255 Whitby, L. E. H. (1938). *The Lancet*, *1*, 1210.
256 Evans, G. M. and Garsford, W. F. (1938). *The Lancet*, *2*, 14.
257 Dale, H. (1958). *Biog. Memoirs F. R. S.*, *4*, 81.

Index